Thin Slab Direct Rolling of
Microalloyed Steels

Thin Slab Direct Rolling of Microalloyed Steels

J.M. Rodriguez-Ibabe

CEIT and Tecnun (University of Navarra)

San Sebastian, Basque Country, Spain

ttp **TRANS TECH PUBLICATIONS LTD**
Switzerland • UK • USA

Trans Tech Publications Ltd
Laubisrutistr. 24
CH-8712 Stafa-Zuerich
Switzerland
http://www.ttp.net

ISBN 0-87849-485-5
ISBN-13: 978-0-87849-485-9

Volume 33 of
Materials Science Foundations
ISSN 1422-3597

Distributed *worldwide by*

Trans Tech Publications Ltd
Laubisrutistr. 24
CH-8712 Stafa-Zuerich
Switzerland

and in the Americas by

Trans Tech Publications Inc.
PO Box 699, May Street
Enfield, NH 03748
USA

Preface

At the beginning of the 90's the Thermomechanical Treatments Group of CEIT started getting involved in steel research projects related to Thin Slab Casting technologies. The initial steps were taken under the direction of the late Prof. Urcola who opened it up to our researchers, as he did in many other topics in which our Department is nowadays involved. Since that period the Thermomechanical Treatment Group of CEIT has been continuously involved in different projects related to near-net-shape technologies, in some cases with public funds and in others collaborating with private industry. The reference point for the latter case being the founding of ACB in 1996, the mini-mill based on CSP technology built close to Bilbao.

Microalloying and thermomechanical processing have been extensively studied by the Group both from the academic point of view and for industrial application. In the first case an important number of doctoral theses with the collaboration of Tecnun (University of Navarra) have been written in the last decade.

As Thin Slab Direct Rolling (TSDR) was confirmed as a technologically possible and economically very interesting industrial route for the production of high added value steel grades, the application of microalloying and thermomechanical processing technologies to TSDR became one of the main research subjects. In this context the contribution of CEIT Research Institute to the field must be considered.

This book attempts to give an overview of the different features of the application of microalloying and thermomechanical processing to thin slab direct rolling technologies. Several courses given during the last two years to personnel from industry and from research groups have been the starting point of the book. Similarly, many of the results and data shown in figures and tables have their origin in different papers and conference contributions published by the Thermomechanical Treatment Group of CEIT. Taking this into account, I wish to thank many of former (A. Linaza, I. San Martin, C. García-Mateo, A. Etxeberria, D. Hernandez, R. Abad, A. I. Fernandez and E. Cotrina) and present (A. Altuna, C. Iparraguirre, M. Arribas, R. Zubialde and B. Pereda) PhD students of the Thermomechanical Treatment Group.

I would like to express my gratitude to my colleagues at CEIT Isabel Gutierrez, Amaia Iza-Mendia, Pello Uranga and Beatriz Lopez for their contribution with interesting discussions and constructive comments to different aspects included in this book. The corrections and changes proposed by Beatriz Lopez are acknowledged.

Finally, I wish to thank my wife, Arantza, for her support and patience during the long preparation period of the manuscript.

Donostia-San Sebastian

February, 2007

Table of Contents

Preface v

1. Introduction 1
    1.1. Different Types of TSDR Industrial Routes 2
    1.2. Economical and Ecological Benefits 5
    1.3. Metallurgical Differences between Conventional
         Rolling and TSDR 6
    1.4 Scope 8

2. **Strength and Toughness: Relationships with Microstructure** 10
    2.1. Introduction 10
    2.2. Strength (in Low Carbon Steels) 12
    2.3. Toughness 16
    2.4. Microstructural Homogeneity and Toughness (in the
         Ductile-Brittle Regime) 18

3. **Microstructural Control during and after Hot Rolling** 30
    3.1. Microstructural Changes during Hot Working 30
         3.1.1. Dynamic Recrystallisation 31
         3.1.2. Static and Metadynamic Recrystallisations 40
         3.1.3. Recrystallised Grain Size 46
         3.1.4. Grain Growth after Recrystallisation 47
    3.2. Conventional Controlled Rolling 48
         3.2.1. Austenite Conditioning 49
         3.2.2. Non Recrystallisation Temperature 53
         3.2.3. Partially Recrystallised Microstructure 60
    3.3. Recrystallisation Controlled Rolling. Effect of Titanium 61
    3.4. Mean Flow Stress 63
    3.5. Phase Transformation during Cooling.
         Precipitation Hardening 65
         3.5.1. Austenite Transformation 66
         3.5.2. Precipitation 73

4. **TSDR: Continuous Casting and Tunnel Furnace** 77
    4.1. Continuous Casting 77
         4.1.1. Micro and Macrosegregations 77
         4.1.2. Transverse Cracks during Continuous Casting 83
         4.1.3. As-Cast Austenite Grain Size 87

4.2. Tunnel Furnace..89

    4.2.1. Effect of Copper and other Tramp Elements in Surface Quality 89

    4.2.2. Microalloying Elements..91

        4.2.2.1. Precipitation of Single Microalloying Elements................................91

        4.2.2.2. Precipitation of Multiple Microalloying Elements................................94

5. TSDR: Rolling of Plain Carbon and Microalloyed Steels 97

  5.1. As-Cast Austenite Refinement..97

    5.1.1. C-Mn and C-Mn-V Steels..102

    5.1.2. NB Microalloyed Steels...103

    5.1.3.   Modelling of Evolution of Grain Size Distributions................................106

  5.2. Austenite Conditioning 112

  5.3. Processing Maps...115

  5.4. Optimisation of Rolling Schedules 121

  5.3. Phase Transformation..125

6. Industrial Applications 127

  6.1. Structural and HSLA steels..127

  6.2. Development of API grades 128

  6.3. Dual phase steels 129

  6.4. Concluding remarks 130

7. References 132

1. INTRODUCTION

Thin slab casting and direct rolling (TSDR) technologies are nowadays one of the most promising processing routes to maintain steel as a leading material in technological applications. Initially, this process was exclusively for the production of mild steels. As industrial experience and knowledge improved, a rapid expansion of the range of products took place with higher strength grades becoming an important part of the overall production. Actually, it is widely accepted as a route to produce high value grades and it can be considered as a technology which has reached a high degree of maturity.

Thin slab direct rolling industrial production started at Nucor in Crawfordsville (USA) in 1989, as a low cost alternative for making hot rolled products. The plant was based on the CSP (Compact Strip Casting) technology developed by SMS. In the CSP process the steel is continuously cast to 50 mm thick "near net shape" slabs that are directly hot rolled into finished strip in a single step. Between the continuous casting and the rolling mill there is a tunnel furnace where the temperature of the thin slab is homogenised.

One of the main characteristics of the process is that the complete casting-rolling process is run in one heat. The productivity of a thin slab caster ranges between 0.9 and 2.0 Mt/year.

After the expansion of CSP technology, several new designs of TSDR technologies have been developed. Some of them are based on different slab thicknesses than the one proposed at the start of the CSP process. In conventional continuous casting routes, slabs are between 200 and 350 mm thick. In near net shape casting of flat products, the following classification, depending on the as-cast thickness has been considered [1]:

— Thin slab: 40 to 125 mm.

— Thick strip: 8 to 25 mm.

— Strip: 1 to 8 mm.

— Thin strip: < 1 mm.

Other authors limit the thin slab term to the range between 40 to 90 mm thickness, while define as medium slabs those thicknesses ranging from 90 to 150 mm [2].

Some of the TSDR plants are supplied with hot metal using integrated plants while in other cases the liquid is provided by electric arc furnaces (EAF) in mini-mills. In the latter, different combinations of scrap and direct reduced iron can be selected depending on the required final product quality.

The introduction of microalloying in thin slab casting and direct rolling technologies started slowly in 1990 with the production of low carbon HSLA (high strength low alloy) steels [3]. It should be noted that the metallurgical peculiarities that differentiate this process from the classical rolling routes, meant that it was necessary to adapt the chemical/processing parameters to the new route so as to achieve similar steel grades as those obtained by traditional routes.

1.1. DIFFERENT TYPES OF TSDR INDUSTRIAL ROUTES

CSP technology was initially designed and developed by SMS, but in the last years a large number of different routes and several improvements have been introduced. Among the various thin slab technologies, the CSP route, with different changes and advances in the latest generations, has gained the greatest acceptance and covers approximately 60-70% of the total TSDR steel production with more than 44 casters distributed worldwide.

All the TSDR technologies have some tools in common, such as pendulum shear, high pressure descalers, hydraulic gauge control and flexible laminar-flow cooling lines enabling the application of different cooling profiles depending on the steel grade. In some plants, liquid core reduction is applied. The mechanical pressure applied during the soft reduction process in the slab caster reduces the central porosity and segregations and leads to an increase in productivity because thicker initial thin slabs are considered.

The main characteristics of different TSDR routes are listed in Table 1. They can be briefly described as follows:

–   Compact Strip Production Process (CSP): one of the main characteristics of CSP technology is the design of a mould with a funnel-shaped top part to provide space to introduce a large diameter submerged nozzle. The thin slab (with a thickness of 50 mm to begin with, although thicker slabs have been included in the latest plants) once cut to the specific length, is transferred to a tunnel furnace approximately 200 m long where the temperature of the slab is homogenised and equalised before going to the finishing mill. Initially the rolling mill at Nucor was composed by 5 stands but there has been a tendency to increase this number including one or two more stands.

–   In-line Strip Production Process (ISP): this technology was developed by Arvedi and Mannesmann Demag. The caster has a parallel mould which provides a thin slab of 60-80 mm including soft reduction technology during continuous casting. The thin slab enters directly in a three stand mill where it is reduced to a pre-determined thickness (between 15 and 25 mm). Once reheated in an inductive heater, the steel is coiled in a twin coil box furnace, called the Cremona furnace. The strip is decoiled and enters into the finishing mill (see Fig. 1). This configuration requires significantly less space than those required when tunnel furnaces are included in the TSDR configuration.

Table 1. Main TSDR routes

Process	Company	Mould type	Thickness (mm)	Furnace	Capacity (Mt/year)	Start-up
CSP	Nucor (USA)	Funnel	50-70	Tunnel	1.8	1989 & 1994
ISP	Arvedi (Italy)	Parallel	60-80	Inductive	0.8	1992
Conroll	Avesta (Sweden)	Parallel	75-125	Pusher type	0.8	1995
QST	Trico (USA)	Parallel	90-105	Tunnel	2.2	1996
fTSC	Algoma (Canada)	Lens-shaped	45-90	Inductive + tunnel	2.0	1997
DSP	Corus (Holland)	Funnel	70	Tunnel	1.3	1999

– Conroll Process (Continuous Casting and Rolling): this was developed by Voest Alpine. The parallel mould provides medium sized slabs of 70 – 80 mm which are transferred to the reheating furnace before entering the finishing mill. One of its initial applications was the production of stainless steels. Ulterior production lines include medium slabs (100-150 mm) linked to 1-2 stand roughing and 5-7 stand finishing mill [4].

– The flexible Thin Slab Casting (fTSC) process developed by Danieli has the following peculiarities: parallel wall mould design which allows peritectic grades, liquid soft reduction and roughing mill followed by a heated transfer table in order to avoid additional temperature loss before entering the finishing mill (Fig. 1).

- The Quality Strip Production: this technique has been developed by Sumitomo Metal Industries and Sumitomo Heavy Industries. It is based on a parallel mould which has the advantage of permitting casting peritectic grades. Slab thickness is close to 90-110 mm. Similarly, some higher casting speed in comparison to the CSP route has been reported [2].

Posterior TSDR developments have included the semi endless rolling process aimed at making ultra light gauge rolling (0.8 mm) easier and also to improve productivity [5]. In this technology, the length of the slab entering the rolling mill is significantly increased and after rolling the strip is cut to the length according to the final coil size. This requires a tunnel furnace longer than in the conventional cases (>300 m) and significantly reduces the problems related to rolling of thin gauges (instabilities in both the head and tail ends of strips, damage on the rolls because thin strip folding, lubrication problems...).

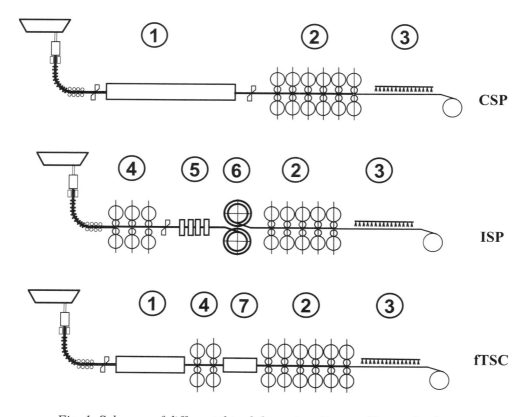

Fig. 1. Schemes of different this slab casting direct rolling technologies.

1: furnace (tunnel); 2: finishing mill; 3: runout table; 4: roughing mill; 5: inductive heater; 6: coil box furnace; 7: heated transfer table.

As observed in Table 1, there is a wide range of different slab thicknesses. Nowadays, the ideal thickness is still open to debate. For example, recently Millet [6] pointed out the convenience to move towards thicker thicknesses than those initially considered taking into account the following aspects related to operation costs, capital and product quality:

- Greater reduction to break down the as-cast structure.

- Less surface area to become defective.

- Greater volume providing better flow conditions.

- Smaller surface/volume ratios reducing yield losses through scaling.

It must be noted that, in many plants, liquid core reduction technology is applied when thicker thin slabs (normally thicker than 60-70 mm) are cast.

1.2. ECONOMICAL AND ECOLOGICAL BENEFITS

TSDR steel production volume is estimated to be close to 15% of the total world hot strip production. In the last few years the opening of new plants has occurred above all in emerging economies, principally in Asia because of the rapid increase in steel demand.

One of the reasons of the spread of TSDR plants is related to both the economical and ecological advantages that can be achieved when compared to conventional routes. If possible economical advantages are considered, the following three aspects need to be evaluated: investment costs, energy consumption and production/labour ratios. Some of these advantages can be summarised as follows:

- Reduction of capital investments associated with tall casters, soaking furnaces and roughing mills that are some of the main characteristics of conventional plants [1,7].

- Energy inputs can be around 25-30% of those required by conventional integrated routes [7]. Fruehan et al. [8] theoretically defined the minimum energy consumption required for steel production evaluating all the production process and in the case of scrap based electric arc furnace routes starting with a 50 mm thin slab, the energy required to produce a 2 mm final gauge thickness was ~ 6 times smaller than that required in the conventional route (blast furnace and reheated 254 mm thick slab).

- Significant cost reductions compared to conventional integrated mills: less than 0.3 man-hour per tonne compared to 1 to 2 man-hours per tonne in the integrated plants [6]. This is also a consequence of the high degree of automation of the new TSDR plants compared to older conventional installations.

- Coiled products are removed from the rolling mill in less than 30 minutes once the thin slab exits from the continuous casting. Contrasting with conventional integrated plants, once the slab has been cooled down the process can take several days (sometimes the slabs can be stored for a long time before reheating and hot rolling; in some situations casting and rolling are done at different locations).

- Short length (from casting to coiling) and less space required for TSDR installation.

- Very flexible process easily adaptable to market requirements.

On the other hand, among the ecological advantages, the following aspects must be considered:

- In those cases where liquid steel is supplied by electric furnaces, the environmental advantages of using recycled scrap need to be considered. Similarly, following this route CO_2 greenhouse gas emissions per tonne of steel produced is reduced in relation to conventional integrated plants (from 0.2 to 0.14 tonne [9]).

- Reduced energy consumptions also have direct ecological implications. For example, during hot rolling the CO_2 emissions are reduced by a factor of 3 from a reheated 250 mm slab to 50 mm direct rolled thin slab [8].

1.3. METALLURGICAL DIFFERENCES BETWEEN CONVENTIONAL ROLLING AND TSDR

Compared to traditional routes, as for example conventional controlled rolling (CCR), TSDR introduces several metallurgical changes. The most significant are the following:

- Smaller centreline segregations during solidification and smaller inclusion size (higher solidification rates and post-solidification cooling).

- Uniform thin slab temperature distribution before entry into the finishing mill which will lead to more consistent final mechanical properties along the hot strip.

- Very coarse as-cast austenite grain size prior to hot rolling. Compared to cold or hot charging, in direct charging there is not austenite-ferrite (during cooling) and ferrite-austenite (during reheating) transformations which can refine the microstructure prior to rolling (see Fig. 2).

- Higher nitrogen and residual element contents (when scrap based EAF routes are considered).

- Smaller total reduction during hot rolling.

- Unavailability for continuous casting in the peritectic carbon range (0.09-0.17 wt pct carbon) in some TSDR technologies (mainly depending on the mould geometry and slab thickness).

All of these peculiarities will have a significant incidence on the behaviour of microalloying elements when thermomechanical processes are considered.

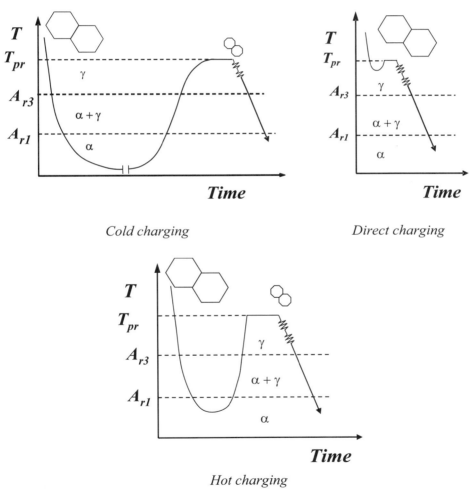

Fig. 2. Comparison between cold, direct and hot charging. In cold and hot charging phase transformation and subsequent grain refinement occur before entering the slab in the rolling mill (T_{pr}: temperature prior to rolling).

Fig. 3 shows the main TSDR stages (considering the configuration corresponding to CSP route) and the corresponding microstructural changes that are involved in microalloying. Similarly, some of the problems that can appear when microalloying is involved, are described. Among others, the loss of microalloying efficiency as a consequence of premature precipitation before rolling, the possible development of surface defects and the presence of microstructural heterogeneities in the final product must be taken into account. To minimise these problems adequate composition/processing parameters must be selected for each application, looking for specific solutions that will be different from those normally applied to conventional routes.

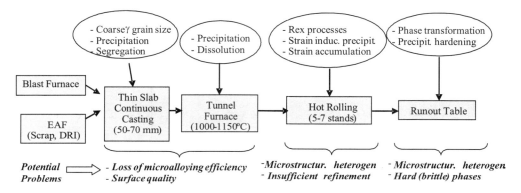

Fig. 3. Scheme showing the metallurgical mechanisms that operate at the different stages of TSDR process as well as and the potential problems related to microalloying technology [10].

Furthermore, there are other two aspects to be considered in the application of microalloying to TSDR technology:

- The optimization of microalloying additions aimed at lowering costs in the production of actual industrial grades.

- The role of microalloying in the development of new grades and/or thicknesses that are now produced in traditional plants for its fabrication by TSDR technology.

1.4.   SCOPE

Taking into account the aforementioned advantages and limitations of TSDR technology, this book aims to provide an approach to the different metallurgical aspects involved in the application of thermomechanical treatments in the TSDR route.

The following chapters analyse some of the special features that affect rolling schedules, microalloying additions and thermomechanical processes. Initially, the main microstructural parameters controlling strength and toughness are described. Hereafter, the microstructural changes affecting hot working will be evaluated in the case of conventional routes, including different classical aspects related to the thermomechanical processing of microalloyed steels.

The metallurgical aspects involved in the continuous casting of thin slabs, as well as the precipitation/dissolution phenomena of microalloying elements before hot rolling will be considered in Chapter 4. In Chapter 5 the applications of the most important basic concepts detailed in Chapters 3 and 4 to the thermomechanical processing of microalloyed steels in TSDR technology are analysed. The final chapter is devoted to an overview of the main industrial applications of this technology.

2. STRENGTH AND TOUGHNESS: RELATIONSHIPS WITH MICROSTRUCTURE

2.1. INTRODUCTION

Thermomechanical processing is the route developed to improve the mechanical behaviour of as-rolled steel products by controlling the hot deformation process. After transformation, the room temperature microstructure is characterised by several parameters which, together with the chemical composition, determine the properties of the steel (Fig. 4). Furthermore, in order to obtain the required properties for each specific application, it is necessary to identify the mechanical properties-microstructure relationships. With the help of this knowledge, it is possible to define the appropriate chemical compositions together with the thermomechanical process parameters that provide the required microstructures.

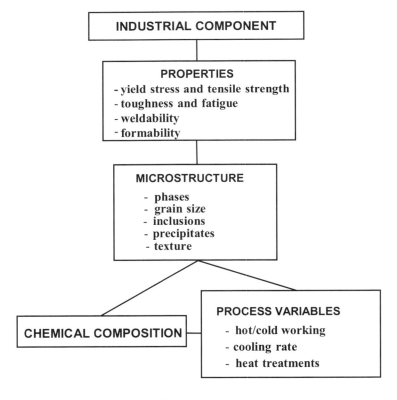

Fig. 4. Scheme correlating final steel properties with composition and process parameters.

Concerning the yield stress, its dependence with microstructural parameters follows the Hall-Petch relationship:

$$\sigma_y = \sigma_o + k_y d^{-0.5} \qquad \text{Eq. 1}$$

where σ_o is the internal stress, d the grain size (ferrite grain size) and k_y a constant. On the other hand, the term σ_o includes the contributions from friction stress (σ_{PN}, Peierls-Nabarro stress), solid solution effects (σ_{ss}), strengthening from precipitation (σ_p) and dislocation hardening (σ_d). Eq. 1 can be rewritten as follows:

$$\sigma_y = \sigma_{PN} + \sigma_{ss} + \sigma_p + \sigma_d + k_y d^{-0.5} \qquad \text{Eq. 2}$$

In the case of non-randomly oriented crystallographies, texture also must be incorporated as an additional term. In the previous expressions a linear additivity approach was taken. However, it has been observed that in some cases this approach can overestimate the yield strength. In order to correct this problem, other addition methods, such as the root mean summation have also been proposed [11].

Concerning the toughness, the dependence between the Charpy impact transition temperature (ITT, defined in some cases as the temperature corresponding to the 50% ductile-brittle behaviour and in other cases, as the temperature at which the absorbed energy is 27 J) and the microstructure is as follows:

$$ITT = f(\sigma_0) - K d^{-0.5} \qquad \text{Eq. 3}$$

Here again the first term can be subdivided to consider the effect of each hardening mechanism:

$$ITT = A + B\sigma_{ss} + C\sigma_p + D\sigma_d + \Phi - K d^{-0.5} \qquad \text{Eq. 4}$$

where A, B, C, D and K are constants and Φ represents the detrimental effect of second phase particles.

Eq. 2 and Eq. 4 clearly show that the refinement of the ferrite grain size is the only procedure that simultaneously enhances both strength and toughness. Other mechanisms, such as solid solution, precipitation hardening and transformation hardening, will always lead to a loss in toughness (measured as an increase of ITT). This relevance of ferrite grain size refinement is the key factor in the development of thermomechanical processes for austenite grain size control in order to achieve refined microstructures at room temperature.

2.2.   STRENGTH (IN LOW CARBON STEELS)

Several semi-empirical relationships have been published to quantify the strength of low carbon C-Mn steels. The more generally accepted equations to quantify the lower yield stress are listed in Table 2.

In these expressions, the contribution of substitutional solute elements (Mn and Si), interstitial solute element (N_f, free nitrogen) and pearlite content (% pearl) are included. In the expression proposed in Ref. [14] the influence of the thickness, including a term with the cooling rate, is considered. In this case the effect of nitrogen is not included because it is assumed that in Al killed steels all the nitrogen is bound in the form of aluminium nitride particles after normalising. On the other hand, equation from Ref. [15] includes the hardening effect of some residual elements.

Table 2. Equations describing (lower) yield strength in low carbon plain steels (in MPa).

Equation	Ref.
$\sigma_y = 88 + 37(\%Mn) + 83(\%Si) + 2918(\%N_f) + 15.1d^{-0.5}$	[12]
$\sigma_y = 105 + 43.1(\%Mn) + 83(\%Si) + 1540(\%N_f) + 15.4d^{-0.5}$ (hot rolled plates with 12 mm thickness)	[13]
$\sigma_y = 68.9 + 12.4(\%Mn) + 101(\%Si) + 1.48(\%pearl) + 16.5d^{-0.5} + 4.93CR^{0.5}$ (steels normalised and cooled at different rates)	[14]
$\sigma_y = 62.6 + 26.1(\%Mn) + 60.2(\%Si) + 759(\%P) + 212.9(\%Cu) + 3286(\%N_f) + {} + 19.7d^{-0.5}$	[15]
Notes: *CR* (cooling rate, °C/min), *d* (ferrite grain size, mm)	

In Fig. 5 the yield stress values calculated with the different equations listed in Table 2 are represented as a function of the ferrite grain size for a selected chemical composition. As can be observed, the differences are smaller than 25 MPa in the range of the grain size considered.

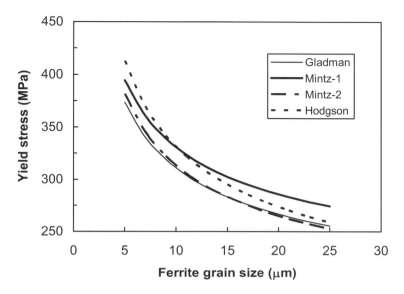

Fig. 5. Evolution of yield stress as a function of ferrite grain size considering equations from Table 2 (1.1% Mn, 0.2% Si, 0.005% P, 0.05% Cu, 50 ppm N_f, 5% pearlite and CR = 60°C/min) (Gladman [12], Mintz-1 [13], Mintz-2 [14] and Hodgson [15]).

When the steels are microalloyed, the term of precipitation hardening also has to be considered. Precipitation strengthening has successfully been described by the Ashby-Orowan model [16]:

$$\sigma_p (MPa) = 10.8 \frac{f^{0.5}}{x} \ln(\frac{x}{6.125 \cdot 10^{-4}}) \qquad \text{Eq. 5}$$

where f is the volume fraction of precipitate and x the mean planar intercept diameter of the precipitate particles (in μm). From the equation it can be seen that the precipitation strengthening rises with an increasing volume fraction and with the refinement in particle size. Taking into account Eq. 5, the relevance of obtaining very fine precipitate sizes is emphasised in Fig. 6. In this context, it must be noted that precipitation strengthening is strongly influenced by the transformation temperature with a significant increase as this temperature decreases.

From a practical point of view the difficulties of using Eq. 5 (mainly due to the need to know the particle diameter) had led to several empirical approaches. In all the cases, the application of these expressions also has some difficulties, as the conditions (mainly cooling rate and temperature at which precipitation happens) significantly affect the size of the particles, and as a consequence the degree of strengthening. One example of this type of empirical relationship is illustrated in Eq. 6 for the case of vanadium steels [17]:

$$\sigma_p (MPa) = 57 \log \dot{T} + 700[V] + 7800[N] + 19 \qquad \text{Eq. 6}$$

where $\dot{T}$ is the cooling rate (°C/s) and *[V]* and *[N]* correspond to the vanadium and nitrogen contents (in wt%) available to precipitate. In this equation the main parameters affecting vanadium precipitation hardening can be identified: cooling rate (refinement of precipitate size), microalloying content (increasing the precipitate volume fraction) and the beneficial effect of high nitrogen contents (as it refines the precipitate size [18]). Other expressions have also been published for the case of V microalloyed steels [19].

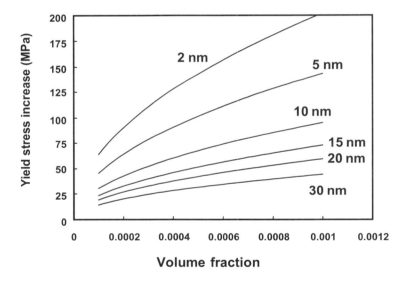

Fig. 6. Effect of volume fraction and particle size in precipitation hardening (in agreement with Eq. 5).

In contrast, referring to niobium the other main microalloying element, more difficulties have been found when determining similar empirical relationships [15]. The more complex precipitation of niobium carbonitrides during and after rolling (including precipitation kinetics) and the significant effect of this element in solid solution during transformation (and as a consequence, modifying the contribution of other strengthening mechanisms) are some of the reasons that can explain these problems [20].

In relation to the dependence of the ultimate tensile strength, *UTS*, on chemical composition and microstructure, several relationships are listed in Table 3. As can be observed, the carbon content increases the UTS (through the increase in pearlite percentage) while it has no significant effect on the majority of the equations published to quantify the yield strength (see Table 2). On the other hand, comparing the expressions in Table 2 and Table 3, it is observed that the contribution of ferrite grain size is significantly higher on the yield strength, nearly twice as much as on the *UTS*.

Table 3. Equations describing UTS in low carbon plain steels (in MPa).

Equation	Ref.
$UTS = 294 + 27.7(\%Mn) + 83.2(\%Si) + 3.85(\%pearl) + 7.7d^{-0.5}$	[21]
$UTS = 164.9 + 634.7(\%C) + 53.6(\%Mn) + 99.7(\%Si) + 651.9(\%P) +$ $+ 472.6(\%Ni) + 339.4(\%N_f) + 11(d)^{-0.5}$	[15]
Notes: CR (cooling rate, °C/min), d (ferrite grain size, mm)	

The aforementioned equations correspond to microstructures formed mainly by ferrite and a small quantity of pearlite. As the strength requirements increase, the contribution of harder phases such as bainite and martensite are necessary. In those situations, as a first approach the law of mixtures can be considered:

$$\sigma_y = \sum f_i \sigma_i \qquad \text{Eq. 7}$$

being f_i the volume fraction of each phase and σ_i its corresponding strength. This implies that the individual contribution of each phase must be determined.

The yield strength of bainite can be quantified with the help of the following expression [22]:

$$\sigma_y(MPa) = 88 + 37(\%Mn) + 83(\%Si) + 2918(\%N_f) + 15.1d_L^{-0.5} + \qquad \text{Eq. 8}$$
$$+ 11(\%Mo) + 1.3(600 - T) + (1172.3 - 2607T)$$

where d_L represents the lath thickness (in mm) and T the transformation temperature (in °C). This equation is very similar to the one proposed for ferrite microstructures in Table 2 but it has two more terms (in addition to the contribution of molybdenum). The first includes the contribution of dislocation density to strengthening and the last one the effect of carbide precipitation (when transformation temperature is lower than 450°C).

In relation to martensite, its strength is affected by the carbon content as follows [23]:

$$\sigma_y (MPa) = 414 + 1724(\%C)^{0.5}$$ Eq. 9

The additivity approach has been successfully applied to predict yield strength in multiphase microstructures, but when work hardening behaviour is required, the complexity of the different strain distributions between the different phases needs to be considered.

2.3. TOUGHNESS

The characterisation of the toughness is often made with the ITT temperature determined from the Charpy curve (see Fig. 7). This temperature has been also related to both the chemical composition and the microstructural parameters. Table 4 shows a list of semi-empirical equations published in the literature relating the ITT temperature to these parameters.

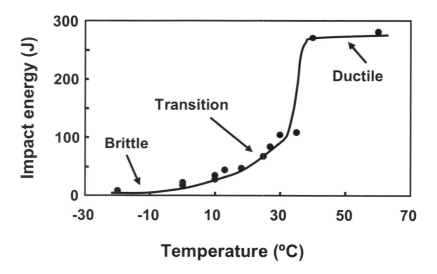

Fig. 7. Charpy curve with the three fracture regimes.

Similarly to the expressions shown to predict the strength, substitutional solute elements (Mn and Si principally), free nitrogen and pearlite content are included in the ITT equations. The important role of refining the ferrite grain size in improvement toughness is well described in the equations.

The combined effect of free nitrogen in yield strength and ITT temperature is shown in Fig. 8 considering the equations in Table 2 (ref. [12]) and Table 4 (ref. [24]). The negative influence of free nitrogen in toughness is clearly evident.

Table 4. ITT temperature equations.

Equation	Ref.
$$ITT = -19 + 44(\%Si) + 700(\%N_f)^{0.5} + 2.2\%pearlite - 11.5(d)^{-0.5}$$ (50% ductile-brittle)	[24]
$$ITT(27J) = 80.1 - 7.41(d)^{-0.5} + 1.4(\%pearl) - 57.2(\%Si) +$$ $$+ 1224(\%S) - 1360(\%P) - 3.7CR^{0.5} - 57.8(\%Mn)$$ *(normalised C-Mn steels cooled at different rates)*	[14]

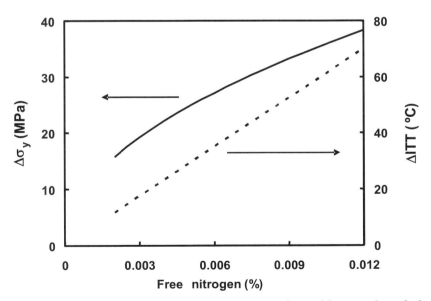

Fig. 8. Combined effect of free nitrogen increasing the yield strength and also significantly modifying the ITT temperature.

Precipitation strengthening affects toughness negatively. In many microalloyed steels a change in transition temperature close to + 0.3°C per 1 MPa increase in yield stress has been observed [25]. For the case of V treated steels Morrison et al. [26] observed that the embrittlement vector is not constant but it increases as precipitation strengthening increases in the range from 0.3 to 0.8 °C/MPa. Nevertheless, it is necessary to take into account the fact that nitride and/or carbonitride precipitation also has a beneficial effect on toughness as they reduce the content of free nitrogen present in the steel (without taking into account other beneficial indirect effects on austenite grain size control and ferrite transformation).

2.4.   MICROSTRUCTURAL HOMOGENEITY AND TOUGHNESS (IN THE DUCTILE-BRITTLE REGIME)

In a significant number of applications toughness is of great engineering significance, together with a good level of strength. In contrast to considerations of strength, the prediction of toughness can become very complex, particularly in the temperature range where brittle-ductile behaviour happens (see Fig. 7).

Some of the reasons for the difficulties of successfully applying the equations listed in Table 4 are related to the following features:

– Presence of mesotexture in the microstructure coming from thermomechanical processes.

– Local microstructural heterogeneities. These can introduce significant dispersions in Charpy tests, as shown in Fig. 9.

– Lack of proper quantification of the detrimental effect of second phase brittle particles.

In this situation, there is a competition between ductile and brittle mechanisms. The control of the micromechanisms intervening in the brittle process can contribute to a significant improvement in the toughness response of the steel. This section will focus on the micromechanisms controlling the cleavage initiation and propagation process.

Brittle cleavage fracture has been attributed to the nucleation of a microcrack followed by its propagation into the surrounding matrix. This process can be divided into three different steps [27, 28]. In the first step (Fig. 10), as a consequence of the tensile stress, a slip-induced microcrack is nucleated in an appropriate microstructural feature which can easily fracture in a brittle manner (a grain boundary carbide, a non-metallic inclusion). The second step is the propagation of the microcrack to the surrounding matrix. This requires the local stress to exceed a certain critical value. The minimum stress σ_{pm} required to propagate a penny-shaped microcrack nucleated at a particle of size a, across the particle-matrix interface, can be defined by the Griffith formula [27]:

$$\sigma_{pm} = \left(\frac{\pi E \gamma_{pm}}{(1-v^2)a} \right)^{1/2} \qquad \text{Eq. 10}$$

where γ_{pm} is the effective surface energy.

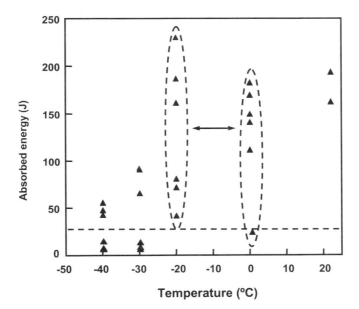

Fig. 9. Example of dispersion in Charpy tests originated by microstructural heterogeneities.

Finally, for the crack progression through the matrix - the third critical step - it is necessary to overcome obstacles present in the matrix in the form of grain boundaries that induce a change in the microcrack plane propagation in order to accommodate the local crystallography. Assuming a D mean grain size, the minimum stress σ_{mm} required for the crack to cross the matrix-matrix (m-m) barrier is defined by the equation:

$$\sigma_{mm} = \left(\frac{\pi E \gamma_{mm}}{(1-v^2)D} \right)^{1/2} \qquad \text{Eq. 11}$$

where γ_{mm} is the corresponding effective surface energy.

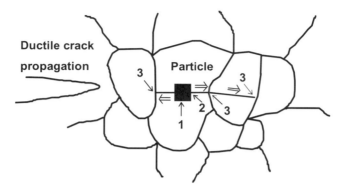

Fig. 10. Scheme of the different steps necessary for cleavage fracture (in ductile-brittle transition). A sharp microcrack nucleates in some microstructural feature (step 1) and propagates across particle-matrix (step 2) and matrix-matrix (step 3) boundaries under the action of a tensile stress

Once the crack has crossed several grains (sometimes one is enough), its length, together with the local stress state, is sufficient for catastrophic propagation to failure. In practice, the process must be dynamic to succeed [29, 30]. Therefore, it is necessary that these three steps described above take place immediately one after the other. If the microcrack stops in one of the steps, a microcrack blunting mechanism occurs, preventing the microcrack from playing a cleavage propagation role.

Fig. 11 shows an example of broken coarse TiN particles where the microcracks stop at the interface and are blunted, acting as nuclei for void formation. Ultimately, the cleavage fracture process could continue when a new microcrack nucleates in an unbroken particle and when the requirements of the three steps are satisfied.

In microalloyed steels, both parameter types (microstructural and energetic) can be modified by adequate microalloying additions and thermomechanical processing, suppressing the cleavage process for a given temperature. Nevertheless, sometimes microalloying itself can contribute to the creation of new microstructural features that may be the origin of microcracks [31]. In this situation matrix microstructure becomes more relevant from the point of view of toughness improvement.

Referring the nucleation of a microcrack, McMahon and Cohen [27] first demonstrated that cleavage fracture in ferrite grains was associated with a broken carbide particle located somewhere in the grain or at the surrounding boundary. Later, in different low carbon weld microstructures inclusion-initiated cleavage fractures were identified too (non-metallic inclusions rich in Ti and Si [32], complex inclusions with Mn, Ti, Si and S [33]). In electric melted low carbon Si-killed structural steels, complex Al, Si and Mn oxides (sometimes together with Ca oxides) have been identified as cleavage nucleation sites [34] and in the case of low and medium carbon Ti microalloyed steels [31, 35] coarse TiN particles (>1 μm). A list of different nucleation particles is reported in Table 5.

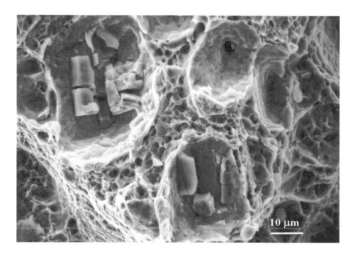

Fig. 11. Broken TiN coarse particles in a ductile matrix [36].

One of the common characteristics of all the inclusion types, reported as being active nucleation sites, is the very strong particle-matrix bonding that they exhibit. This bonding is a consequence of the mismatch between the non-metallic particle and the matrix. Taking into account the work of Brooksbank and Andrews about the residual stresses associated with different thermal expansion coefficients [37], the most deleterious particles are calcium aluminates, alumina, TiN and some silicates.

From the point of view of size, in agreement with Eq. 10, the coarsest particles would be the most deleterious ones. Nevertheless, the influence of the size of the particles can only be considered when taking into account the availability for the microcracks, thus produced to propagate across the matrix. Finally, it is obvious that the volume fraction of these brittle particles will have repercussions on the toughness; nevertheless, this influence is difficult to quantify.

Once the microcrack has been nucleated, it is necessary to consider the second step: its successful brittle propagation into the matrix. According to Eq. 10, the minimum stress σ_{pm} required depends on the microcrack size and on the γ_{pm} effective surface energy. A value of $\gamma_{pm} = 7$ J/m^2 has been proposed for a wide range of steels [32, 41]. This energy is usually considered to remain constant (or change only very slightly) with temperature.

Table 5. Second phase particles reported as being active on the nucleation of microcracks.

Cracking of a microstructural feature		
Feature	**Steel**	**Ref.**
Carbides	Ferritic steels, bainitic steels quenched and tempered steels	[27, 38]
Complex inclusions	Low carbon welds, eutectoid steels,	[29, 38]
Al_2O_3-MnO-SiO$_2$, Al_2O_3-MnO-SiO$_2$-CaO	Si killed carbon and V microalloyed steels	[34]
TiN particles (>1µm)	Ti microalloyed steels	[31, 35, 39]
Martensite-austenite films	HAZ, HSLA steels	[40]
Microcracks emanating from voids		
MnS elongated inclusions, ductile voids in matrix	Pearlitic steels, bainitic steels, acicular ferrite steels	[40 - 42]

In step 2 the misorientation angle, β, between the normal to the first formed cleavage facet and the direction of the applied tensile stress is another relevant parameter (see Fig. 12). For microcrack propagation from the particle to the matrix it is necessary for at least one of the normals to the {100} crystallographic planes of the grains surrounding the broken particle to be close to the tensile stress component application direction. In practice, β is found to be relatively low [43].

Step 2 of the cleavage process is considered to be the controlling step at very low temperatures (the zone corresponding to the completely brittle behaviour in the lower plateau of the Charpy curve) [44, 45]. In contrast, in the ductile-brittle regime, matrix-matrix boundaries become, in the majority of situations, the most relevant step in the cleavage propagation process.

Once the first cleavage facet is formed the successful propagation of the brittle microcrack depends on the trespassing of high angle matrix-matrix boundaries. In this situation, as happens with step 2, the two parameters that must be taken into account are the effective surface energy, γ_{mm} , and the grain size, D.

Following the approaches proposed by Linaza et al. [44, 45], San Martin et al. [46] estimated the variation of γ_{mm} with temperature for a vanadium microalloyed steel. As indicated in Fig. 13, there is a significant increase in γ_{mm} with temperature. This means that as temperature increases, matrix-matrix boundaries become increasingly difficult to trespass and as a consequence, microcracks nucleated at brittle non-metallic inclusions can be arrested in step 3. It is worth emphasising that this change (related to the plastic contribution to γ_{mm}) is responsible for the relevant role of step 3 in the ductile-brittle transition, while for very low temperatures, into the brittle range, the cleavage process is controlled by step 2.

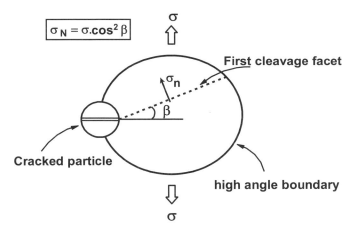

Fig. 12. Misorientation angle β between the perpendicular to the first formed cleavage facet and the direction of the applied tensile stress.

Assuming that σ_F, cleavage fracture stress, and γ_{pm}, the effective surface energy, are independent of the temperature, the explanation for the improved toughness by microstructural refinement could be because of the influence of temperature on γ_{mm} energy, acting as the controlling parameter for the brittle processes in the transition regime. This means that, considering γ_{pm} as a constant, the microstructural refinement would provide better toughness at any temperature. On the contrary, if γ_{mm} increases with the temperature, there will be a temperature range in which the matrix-matrix boundary could become the controlling factor of the cleavage fracture.

Based on the aforementioned arguments, Fig. 14 aids the interpretation of the results [45]. Since as can be seen in the figure γ_{pm} remains constant with temperature and γ_{mm} increases with it, it is possible to define a critical "grain size" value for each temperature. Since control of cleavage fracture may be by step 2 or step 3, when:

$$\text{step } 2 : \sigma_{pm} > \sigma_{mm} \rightarrow \frac{\gamma_{pm}}{a} > \frac{\gamma_{mm}}{D} \rightarrow \frac{D}{a} > \frac{\gamma_{mm}}{\gamma_{pm}} \qquad \text{Eq. 12}$$

$$\text{step } 3 : \sigma_{pm} < \sigma_{mm} \rightarrow \frac{\gamma_{pm}}{a} < \frac{\gamma_{mm}}{D} \rightarrow \frac{D}{a} < \frac{\gamma_{mm}}{\gamma_{pm}} \qquad \text{Eq. 13}$$

are true respectively, so for a value of $T = T_1$ (Fig. 14) there exists a critical value of γ_{mm}/γ_{pm} which turns out to be equal to $(D/a)_1$. Consequently, for grain sizes lower than D_1, cleavage fracture will be controlled by step 3 and the refinement of the microstructure will improve the toughness because of the arrest of the cracks at matrix-matrix boundaries. In contrast, if the grain size is coarser than D_1, step 2 will control cleavage fracture and matrix-matrix boundaries will not be able to stop the brittle processes.

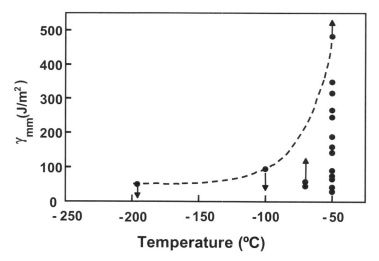

Fig. 13. Variation of γ_{mm} energy with temperature. For temperatures lower than -100°C the data correspond to upper limits and for higher temperatures they are lower limits of the real γ_{mm} value [46].

From the microstructural point of view some parameters that can be considered to define the m-m barriers are listed in Table 6. In the case of heat treated (or conventionally hot rolled) low carbon steels with predominantly ferrite microstructures, the ferrite grain size corresponds to the cleavage facet size. In ferrite-pearlite medium C steels, the microstructural barriers that can stop brittle microcracks are "ferrite units" composed of ferrite plus pearlite with the same ferrite orientation [31, 46]. In a similar way, in pearlitic microstructures the unit with the same ferrite crystallographic orientation is the relevant parameter [47 - 49]. In bainitic steels the packet size is the m-m barrier [50, 51] and in acicular ferrite steels the plate size [52, 53]. Finally, for the case of thermomechanically processed low carbon ferritic steels the relevance of mesotexture in the ductile-brittle region has been emphasised [54]. In this context, the mesotexture also appears as the key factor for the definition of the microstructural unit controlling cleavage crack propagation in more complex microstructures. In general, a minimum crystallographic misorientation of 12-15° has been considered as appropriate to define the microstructural unit that controls cleavage propagation [53, 54].

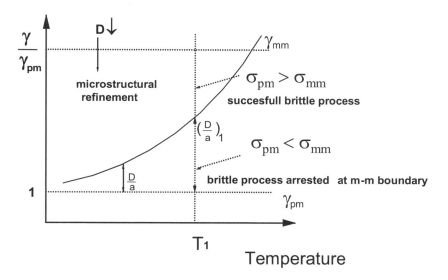

Fig. 14. Scheme of the relationship between γ_{pm} and γ_{mm} energetic barriers and the corresponding microstructural feature controlling cleavage propagation as a function of temperature [45].

Independently of the control of the volume fraction and the nature of particles and the features able to nucleate microcracks (steps 1 and 2 of brittle propagation), the control of the matrix-matrix boundaries implies that a fine and homogeneous microstructure in the whole volume of the material must be achieved. Due to the weakest link character of the brittle process, the volume fraction of coarser units is more relevant than the overall mean unit size. This condition becomes the key factor in many industrial applications where a fraction of coarser grain sizes, not relevant from the point of view of strength, significantly impairs toughness values in the ductile-brittle regime moving the ITT temperature to higher values.

Table 6. Microstructural units involved in the 3rd cleavage propagation step.

Unit	Matrix	Ref.
Ferrite grain size	Ferrite (heat treated)	[27]
"Effective grain" size	Ferrite (thermomechanically processed)	[54]
Ferrite unit (α + pearlite with the same α orientation)	Ferrite-pearlite	[31, 43]
Unit with same ferrite orientation	Pearlite	[47 - 49]
Packet size	Bainite	[50, 51]
Plate size	Acicular ferrite	[52, 53]

Fig. 15 shows an example where the main fracture initiation site, in a bainitic Ti-V microalloyed steel, exhibits a cluster of coarse facets [55]. In the same figure the facet size distribution together with the size of the first cleavage facets (of several specimens) is illustrated, indicating that 70% of the facets are smaller than those responsible for brittle fracture initiation.

Fig. 16 shows the tensile test values, the mean ferrite grain size and the ferrite grain size distribution of a Ti-Nb microalloyed steel with two different thermomechanical schedules [56]. In both cases the strength is very similar, but there are main differences in the ferrite grain size distribution.

The microstructure with D_α = 9.7 μm shows a spread of the distribution to larger grain sizes, with a significant fraction of grains coarser than 20 μm. In both cases there are TiN coarse particles originated during the solidification process. Nevertheless, for the D_α = 8.2 μm case, as can be observed inside the voids (see Fig. 16-b), the microcracks initiated in a brittle manner on coarse

TiN particles (step 1 of cleavage process) have been stopped at the particle-matrix interface (step 2), giving rise to ductile fracture modes with void nucleations. In contrast, the coarser and more heterogeneous microstructure present in the $D_\alpha = 9.7$ µm case has promoted the propagation of the microcracks to the ferrite matrix (step 2) leading to a catastrophic propagation to failure (step 3). Thus, comparing the two situations, an improvement in toughness can be achieved only if a well-refined microstructure is obtained.

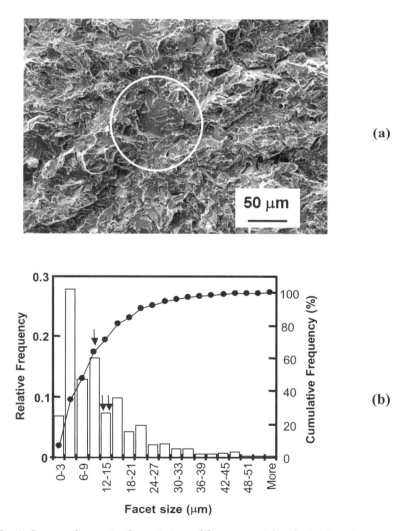

(a)

(b)

Fig. 15. a) Coarse facets in the origins of fracture with a bainitic microstructure and b) facet size histogram with the corresponding distribution function. Arrows denote the facet sizes of the "first cleavage facet" [55].

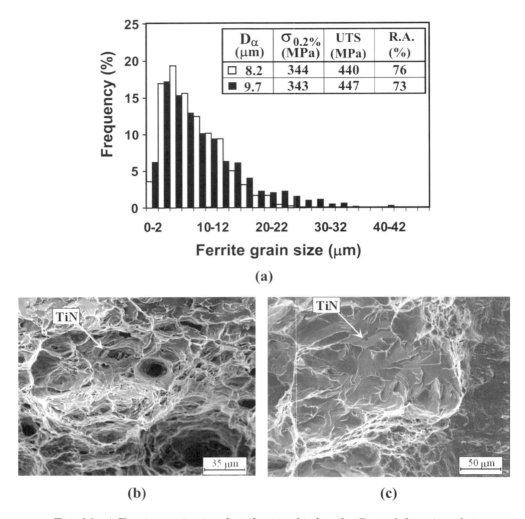

Fig. 16. a) Ferrite grain size distribution, b) ductile (D_α = 8.2 μm) and c)
ductile/brittle behaviour (D_α = 9.7 μm) of a Nb-Ti steel tested at -20°C [56].

Summarising, for applications where high toughness requirements are
necessary, several recommendations must be taken into account:

—   Reducing as much as possible the presence of harmful particles that can
    act as nucleation sites for cracks. In this case the nature of particles as well
    as their size must be controlled.

—   Refinement and general homogenisation of the microstructure to stop the
    propagation of the cracks formed.

As will be shown in Chapter 5, this latter requirement becomes very relevant when the appropriate thermomechanical schedules need to be designed in the TSDR route.

3. MICROSTRUCTURAL CONTROL DURING AND AFTER HOT ROLLING

3.1. MICROSTRUCTURAL CHANGES DURING HOT WORKING

During the hot rolling of flat products, each pass is characterised by the applied strain, strain rate and temperature, together with the interpass time. Assuming some basic considerations, both strain and strain rate can be described by the expressions shown in Fig. 17. In this situation, for a given initial microstructure, characterised by its mean austenite grain size, D_o, the strength of the steel will increase when the applied strain rate is higher and when the temperature decreases. On the other hand, the deformation in hot working conditions brings about dynamic structural changes which leave the steel in an unstable state. At these conditions, recovery and recrystallisation will take place and, if there is enough time, subsequent grain growth as shown in the scheme in Fig. 18.

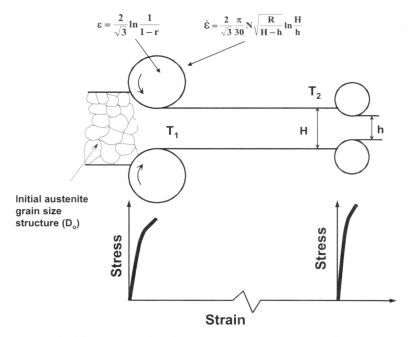

$$\varepsilon = \frac{2}{\sqrt{3}} \ln\frac{1}{1-r} \qquad \dot{\varepsilon} = \frac{2}{\sqrt{3}} \frac{\pi}{30} N \sqrt{\frac{R}{H-h}} \ln\frac{H}{h}$$

Fig. 17. Scheme showing the main parameters of a rolling pass.

The analysis of the austenite modifications during hot rolling requires the development of relationships able to describe the microstructural changes that take place in the austenite during the process. These include:

- Dynamic recrystallisation.

- Static and metadynamic recrystallisation during interpass time intervals.

- Grain growth during reheating and after recrystallisation.

- Interaction of recrystallisation with precipitation.

- Transformation to ferrite and/or other phases (final grain size control).

- Precipitation hardening.

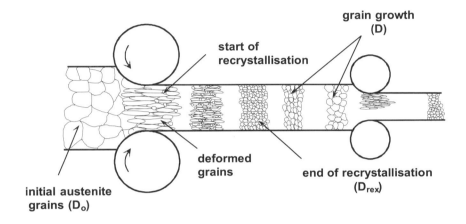

Fig. 18. Scheme identifying the different microstructural changes happening during hot working.

The application of models is one of the most appropriate procedures to analyse these changes. In the last few years much more attention has been paid to developing mathematical models that predict the final microstructure of a product depending on the processing conditions. The application of these models allows deformation sequences to be optimised in order to obtain a good combination of mechanical properties in the as-rolled materials. Most of these models are based on empirical equations developed to describe the microstructural events that occur in the austenite during hot working and cooling. In the following sections, an analysis of the main microstructural changes that take place during hot working and the bases of the thermomechanical processing will be described.

3.1.1. DYNAMIC RECRYSTALLISATION

During the hot working of austenite microstructures a competition between hardening and softening mechanisms occurs. For a given temperature and strain rate the stress-strain curve takes the form seen in Fig. 19. The initial rapid rise in stress is associated with an increase in dislocation density as a result of the work hardening. In the case of low stacking fault energy materials, like austenite, the dynamic recovery process is very slow and the dislocation density can attain a sufficiently high value to allow the nucleation of new recrystallised grains during deformation. This softening mechanism, called dynamic recrystallisation (*DRX*), starts once a critical strain, ε_c, is exceeded and leads to the elimination of a large number of dislocations by means of the migration of high angle boundaries [57]. The softening causes the flow stress to go through a maximum (denoted by the peak strain, ε_p) before falling to a steady state (ε_{ss}).

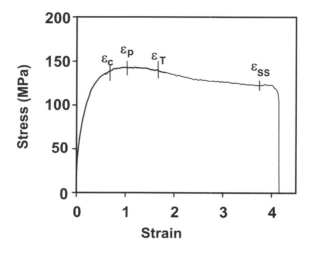

Fig. 19. *Stress-strain curve typical of dynamic recrystallisation with the main singular strains.*

Consequently, dynamic recrystallisation occurs during deformation and both nucleation and grain growth take place while the strain is being applied. The new grains nucleate preferentially at existing grain boundaries of the microstructure. As the grains grow, the applied strain builds up the dislocation density inside them and gradually reduces the driving force for boundary migration until growth becomes completely restricted. At these conditions, the recrystallisation proceeds by the nucleation of successive new grains at the interface between the previously recrystallised grains and the remaining part of the original grain, through a "necklace" process [58], as shown in the illustration of Fig. 20. These grains are in turn deformed until they once again reach the critical strain required to undergo recrystallisation, resulting in a "cascade" of nucleation and limited growth events. This continuous sequence of concurrent deformation and recrystallisation leads to a steady state, characterised by the maintenance of a structure of approximately equiaxed grains of a constant mean grain size, D_{dyn}, and stress, σ_{ss}, independent of strain. The strain value at which this steady state is achieved is denoted as ε_{ss}. An example of the evolution of the dynamically recrystallised microstructure with the applied strain is shown in Fig. 21 for the case of a Nb microalloyed steel with a very coarse initial grain size. At these conditions the new dynamically recrystallised grains are easily identified.

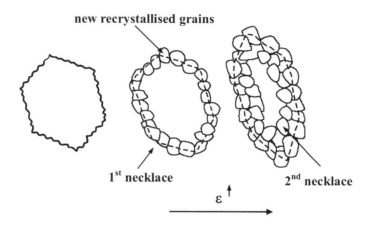

Fig. 20. Scheme illustrating the dynamic recrystallisation progression with applied strain.

Knowledge of the value of the critical strain for the onset of *DRX*, ε_c, is required for the evaluation of the softening mechanisms in hot rolling and their influence on subsequent microstructure evolution. This critical strain has been related to the peak strain, ε_p, i.e. the strain corresponding to the maximum stress in the stress-strain curve (see Fig. 19). Usually the peak strain depends on the initial grain size (D_o) and the Zener-Hollomon parameter (Z) by equations of the type [59]:

$$\varepsilon_p = BD_o^m Z^p \qquad \text{Eq. 14}$$

$$Z = \dot{\varepsilon}\exp\left(\frac{Q_{def}}{RT}\right) = A(\sinh \alpha\sigma)^n \qquad \text{Eq. 15}$$

where $\dot{\varepsilon}$ is the strain rate, Q_{def} the apparent activation energy for deformation, R the gas constant (8.31 J/K.mol) and T the absolute temperature. The activation energy Q_{def}, the coefficients A, B and α, and the exponents m, p and n are dependent on the material [59 - 63]. Different values found in the literature for the case of plain C-Mn steels are listed in Table 7.

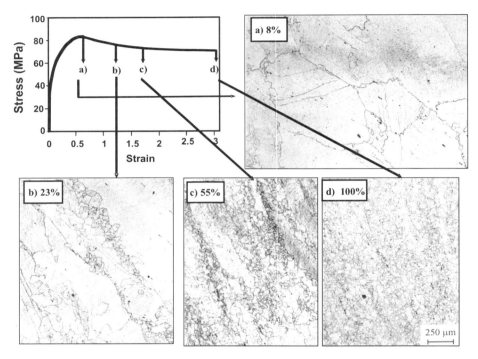

Fig. 21. Dynamic recrystallisation evolution in a Nb microalloyed steel deformed at 1100°C in a laboratory torsion test at a strain rate of 0.2 s^{-1} with $D_o = 800$ μm. Metallographically measured values corresponding to the recrystallised fraction are shown in each micrograph [64].

Eq. 14 denotes that the initial grain size is an important variable for the onset of dynamic recrystallisation: a fine grain size promotes dynamic recrystallisation, while a coarse one retards it, as can be observed in the example of Fig. 22 [65]. This feature is relevant in the case of as-cast coarse grains during the first passes in direct hot rolling.

Microalloyed elements in solid solution in the austenite may retard dynamic recrystallisation [66] and consequently, affect the peak strain value. In fact in Fig. 22, the effect of the amount of dissolved Nb and Ti is also included, since the different grain sizes were obtained by applying different reheating conditions [65]. It must be taken into account that increasing the reheating temperature increases the degree of dissolution of microalloying elements and the grain size.

Table 7. Values of parameters of Eq. 14 and Eq. 15 corresponding to the peak strain for plain C-Mn steels (D_o in μm).

B	m	p	Q_{def} (kJ/mol)	Ref.
4.9×10^{-4}	0.5	0.15	312	[59]
6.97×10^{-4}	0.3	0.17	312	[60]
1.32×10^{-2}	0.174	0.165	293	[61]

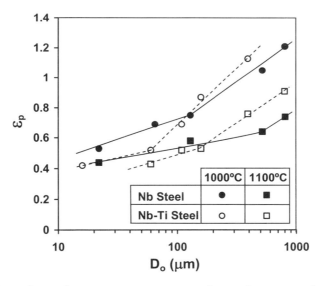

Fig. 22. Effect of initial austenite grain size on the peak strain ε_p for Nb and Nb-Ti microalloyed steels at two different deformation temperatures ($\dot{\varepsilon} = 1\ s^{-1}$) [65].

In the case of Nb microalloyed steels, Minami et al. [67] quantified the effect of Nb in solution multiplying the proportionality constant B of Eq. 14 by a term dependent on the Nb content:

$$\varepsilon_p = B \frac{1 + 20[C]}{1.78} D_o^m Z^p \qquad \text{Eq. 16}$$

where [C] is the Nb concentration in solution in weight percentage.

For Nb-Ti microalloyed steels, the same approach was used but considering an expression for [C] corrected by the amount of Ti present in solid solution as follows [65]:

$$[C] = [Nb] + 0.02\,[Ti]$$
Eq. 17

[Nb] and [Ti] being the respective niobium and titanium concentrations in solution in wt. % in austenite. This equation denotes that Ti in solid solution has an effect significantly smaller than Nb in retarding *DRX*.

From the above, the data of ε_p shown in Fig. 22 were corrected to take into account the effect of solute drag. A solute drag corrected peak strain was defined as:

$$\varepsilon_p^* = \frac{\varepsilon_p}{1 + 20[C]}$$
Eq. 18

The relationship between the correct peak strain and the initial grain size for a given level of *Z* is shown in Fig. 23. From the figure it is observed that the points can be fairly fitted to just one straight line for the large range of grain sizes studied (16 – 805 μm) for both steels [65].

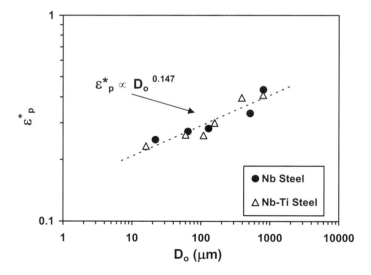

Fig. 23. Evolution of the solute drag corrected peak strain, ε_p^*, with initial grain size for Nb and Nb-Ti microalloyed steels (Z = 10^{12} s^{-1}) [65].

Referring to the rest of the microalloying elements, their influence would be significantly lower, in agreement with Jonas [66], as shown in Fig. 24. This figure illustrates the effect of 0.1% single microalloy additions (in weight percent) of V, Mo, Ti and Nb on the solute retardation parameter (*SRP*), compared to what was observed on a plain C-Mn steel [66, 68]. This parameter is defined as:

$$\mathbf{SRP} = \mathbf{log}(\frac{\mathbf{t_x}}{\mathbf{t_{ref}}}) \cdot \frac{\mathbf{0.1}}{\mathbf{wt.\%x}} \cdot \mathbf{100\%}$$
Eq. 19

where t_x is the time to the peak strain for the steel containing the element x and t_{ref} is the equivalent time for the case of a plain carbon steel.

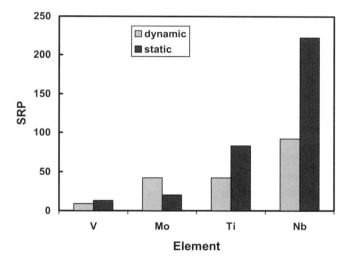

Fig. 24. Effect of V, Mo, Ti and Nb on the solute retardation parameter (SRP) for dynamic and static recrystallisations [66, 68].

Fig. 24 shows that Nb is the most effective element in retarding recrystallisation followed by Ti, Mo and V in decreasing order of effectiveness. Concerning the situations corresponding to other combinations of microalloying elements, Akben and Jonas [68] observed relative small differences between the peak strain values measured in Nb microalloyed steels and those determined for Nb-Mo and Nb-Mo-V steels. Nevertheless, some recent works [69] on Nb-Mo steels with Mo contents between 0.15-0.30% have shown a stronger solute effect of Mo than that predicted by the above authors.

On the other hand, Medina and Hernandez [62] have proposed that in Eq. 14 it would be better to express ε_p as a function of Z/A instead of a unique dependence on Z, A being the material dependent coefficient of Eq. 15. Following this approach, they also identified that Mo contributes to increase the peak strain in a greater degree than V and Ti.

Different equations proposed for determining the peak strain in the case of microalloyed steels are compiled in Table 8.

Table 8. Equations describing the peak strain for microalloyed steels (D_o in μm).

Steel	Equation	Ref.
Nb	$\varepsilon_p = 2.8 \times 10^{-4} \dfrac{\{1 + 20[\text{Nb}]\}}{1.78} D_0^{0.5} Z^{0.17}$ $Q = 375 \, \text{kJ/mol}$	[67]
Nb and Nb-Ti	$\varepsilon_p = 3.7 \times 10^{-3} \dfrac{\{1 + 20([\text{Nb}] + 0.02[\text{Ti}])\}}{1.78} D_0^{0.147} Z^{0.155}$ $Q = 341 \, \text{kJ/mol}$	[65]

As mentioned above, once ε_p is defined, it is possible to know the value of the critical strain ε_c. The relationship between ε_c and ε_p is complex but it has been suggested that $\varepsilon_c = k\varepsilon_p$, where k is a constant, whose reported values range from 0.5 to 0.87 in the literature [15, 59, 70]. In general, the constant is about 0.8 in plain C-Mn steels and it drops to 0.5-0.6 when microalloying elements as Nb are present.

Siciliano and Jonas [70] proposed that, for Nb microalloyed steels, the constant k was dependent on the chemical composition:

$$k = 0.8 - 13\text{Nb}_{\text{eff}} + 112\text{Nb}_{\text{eff}}^2$$

$$\text{with } \text{Nb}_{\text{eff}} = \text{Nb} - \frac{\text{Mn}}{120} + \frac{\text{Si}}{94}$$

Eq. 20

These equations are thought to be applicable over the following composition ranges: 0.010 to 0.058% Nb, 0.35 to 1.33% Mn and 0.01 to 0.23% Si. The opposing influences of Mn and Si are due to their different effects on Nb diffusitivity in austenite [67] and as a consequence, on Nb solute drag at austenite grain boundaries. This means that high levels of Mn and low levels of Si favour the occurrence of dynamic recrystallisation (that is, lower ε_c values).

The dynamically recrystallised grain size depends only on the Z parameter, being completely independent on the initial microstructure [58, 59]. This dependence follows a power relationship in the form:

$$\mathbf{D_{dyn} = BZ^{-n}} \qquad\qquad \text{Eq. 21}$$

where B and n are two constants dependent on the chemical composition. The exponent value n reported in the literature ranges from 0.11 to 0.35 [59, 71]. In Fig. 25 the dependence of D_{dyn} with Z is shown for the case of a Nb microalloyed steel [65].

At high Z values, the higher dislocation accumulation results in a higher strain hardening of the new recrystallised grains. As a consequence, a reduction in the growing driving force occurs, leading to smaller dynamically recrystallised grain sizes.

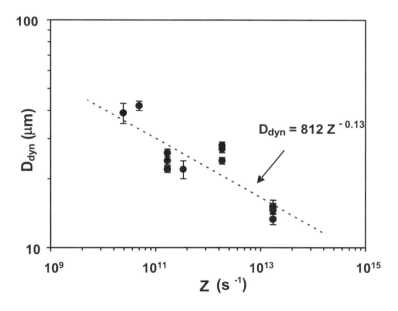

Fig. 25. Evolution of dynamically recrystallised grain size with Zener-Hollomon parameter [65].

3.1.2. STATIC AND METADYNAMIC RECRYSTALLISATIONS

After deformation, during the interpass time intervals, softening of the austenite can occur by static recovery and recrystallisation. Recovery in the austenite is very limited, recrystallisation being the most important softening mechanism present. Depending on the pass-strain there should be a difference made between static recrystallisation (SRX), when $\varepsilon < \varepsilon_c$, and metadynamic recrystallisation $(MDRX)$, when $\varepsilon > \varepsilon_c$. Both softening mechanisms differ in several ways.

Nevertheless, at this point it should be noted that some recent works disagree with the criteria of using the critical strain, ε_c, to distinguish both working ranges and propose that the characteristics of $MDRX$ process are only achieved after a minimum strain, denoted as the transition strain, ε_T, is reached (see Fig. 19). Depending on deformation and microstructural conditions, this strain could be significantly larger than ε_c [72, 73].

The evolution of the statically recrystallised fraction with time can be described by the Avrami equation:

$$X = 1 - \exp\left(-0.693\left[\frac{t}{t_{0.5}}\right]^n\right)$$
Eq. 22

where X is the recrystallised fraction after time t, $t_{0.5}$ is defined as the time required to reach a 50% recrystallisation, and n is the Avrami exponent. An example of the fit of several experimental results to Eq. 22 is shown in Fig. 26 for the case of a Nb microalloyed steel [74]. In the figure the influence of initial austenite grain size (D_o) and applied strain are shown.

The values of the Avrami exponent n reported in the literature range between 1 and 2 [15, 74 - 77]. Laasraoui and Jonas proposed a value of $n \approx 1$ for C-Mn and Nb microalloyed steels [76]. Medina and Mancilla [78] suggested a slight dependence on the temperature with values of n between 0.62 and 1.50 (a similar effect is reported in ref. [77] for the case of Ti microalloyed steels). On the other hand, values of n close to 2 have also been reported in the bibliography [59, 79].

Static recrystallisation kinetics in C-Mn and microalloyed steels have been extensively investigated and several regression equations have been proposed. The influence of prior austenite grain size (D_o), applied equivalent strain, deformation temperature and strain rate on the kinetics of static recrystallisation are conveniently expressed in terms of the time required to induce some specified recrystallised fraction, say $X = 0.5$.

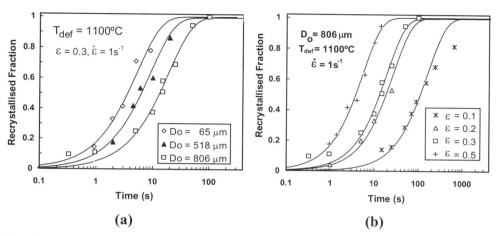

Fig. 26. Evolution of recrystallised fraction with time for a 0.035% Nb microalloyed steel: a) influence of initial austenite grain size and b) effect of applied strain [74].

The general expression that takes into account the influence of the different parameters that intervene on the value of $t_{0.5}$ is [59]:

$$t_{0.5srx} = A \varepsilon^{-p} \dot{\varepsilon}^{-q} D_0^m \exp(Q/RT) \qquad \text{Eq. 23}$$

where m, p and q are constants and Q is the activation energy for static recrystallisation.

The impact of the strain in Eq. 23 can be clearly understood taking into account the fact that the difference in dislocation density between deformed and undeformed areas constitutes the driving force for recrystallisation. Consequently, increasing the strain produces an increment of the dislocation density providing a higher driving force for recrystallisation, which results in a decrease in the recrystallisation time. A power law dependence of the type $t_{0.5srx} \propto \varepsilon^{-p}$, with p being a constant taking a value between 2 and 4, is usually found in the literature [59, 61, 80, 81].

The effect of strain rate is less than the influence of the other variables, but according to several authors not insignificant, leading to a decrease in the recrystallisation time as the strain rate increases. A typical dependence of the type $t_{0.5srx} \propto \dot{\varepsilon}^{-q}$, with q a constant taking different values for different steel grades, is usually reported.

Regarding the material parameters, the effect of the grain size must be considered first. Recrystallised grains nucleate mainly on austenite grain boundaries. When the grain size decreases the specific grain boundary area increases which provides more sites for the nucleation of new grains and consequently lowers the time required for recrystallisation. A dependence of $t_{0.5srx} \propto D_o{}^m$, is usually observed. An exponent *m=2* has often been reported in the literature [59]; however, according to other authors, a linear dependence, i.e. *m=1*, agrees better with experimental data [74, 82], as shown in Fig. 27. This can be more relevant for the case of thin slab casting where very coarse as-cast grain sizes will be present at the entry of the rolling mill.

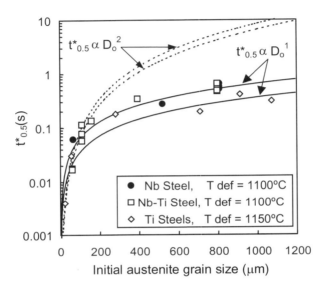

Fig. 27. Effect of initial grain size on the time for 50% recrystallisation in different microalloyed steels [74].

In microalloyed steels another factor that may influence the recrystallisation kinetics is the solute drag effect exerted by the microalloying elements dissolved in the austenite during reheating. It is well known that the microalloying elements in solution can retard *SRX*, even though they are less effective than when precipitated [66].

Dutta and Sellars [15] proposed an equation for the 50% recrystallisation time in Nb-bearing steels which took into account the effect of Nb as follows:

$$t_{0.5srx} \propto \exp\left(\left\{\left[\frac{275000}{T}\right]-185\right\}\cdot[C]\right)$$ Eq. 24

where *[C]* is the Nb concentration in solution in wt.% and *T* is the deformation temperature.

The solute effect of different microalloying elements on static recrystallisation was studied by Akben and Jonas [68]. These authors, with the help of the solute retardation parameter for static recrystallisation (*SRP*), quantified the delay produced in recrystallisation time by the addition of 0.1% of the different alloy elements with respect to a C-Mn base steel. They found that niobium was the most potent element in retarding static recrystallisation, followed by Ti, Mo and V in decreasing order of effectiveness, as shown in Fig. 24. For niobium this parameter reached a value of SRP = 222, SRP = 83 for titanium and SRP = 13 for vanadium. Using these parameters to obtain the same drag effect as that produced by 0.1 wt. % Ti, an amount of 0.0374 wt.% Nb would be sufficient and similarly, an amount of 0.00585% Nb for the case of 0.1% V. Taking this into account, Eq. 24 might also be applied to other microalloying elements if the solute concentration of each of these elements is modified by the corresponding multiplying factor. Therefore:

$$[C] = [Nb] + 0.374[Ti] + 0.0585[V] \qquad \text{Eq. 25}$$

This type of approach has been successfully applied to vanadium [83] and Nb-Ti [74] microalloyed steels.

Several equations proposed in the bibliography for the calculation of recrystallisation time are listed in Table 9.

Once a significant dynamic recrystallisation has occurred there is a post-dynamic recrystallisation process between passes, termed metadynamic recrystallisation. This recrystallisation removes the previous dynamically recrystallised microstructure.

The kinetics of metadynamic recrystallisation can also be described by the Avrami equation (Eq. 22), in this case the values of the Avrami exponent are in general close to 1 [73, 84]. Fig. 28 illustrates the influence of strain rate on the *MDRX* kinetics for the case of a Nb microalloyed steel and the corresponding Avrami plots of the experimental data [73].

The time to reach 50% of metadynamic recrystallisation, $t_{0.5mdrx}$, depends on the deformation parameters as follows:

$$t_{0.5mdrx} = B \cdot Z^{-r} \cdot \exp\left(\frac{Q_{mdrx}}{RT}\right) = B\dot{\varepsilon}^{-r} \exp\left(\frac{Q_{app}}{RT}\right) \qquad \text{Eq. 26}$$

where Q_{app} is an apparent activation energy, which is a function of both the activation energy for metadynamic recrystallisation, Q_{mdrx}, and the activation energy of deformation, Q_{def}. Several examples of empirical equations obtained for C-Mn and Nb microalloyed steels are shown in Table 10.

Table 9. Equations describing time (in s) for 50% static recrystallisation (D_o in μm).

Steel	Equation	Ref.
C-Mn	$t_{0.5srx} = 2.5 \cdot 10^{-19} \varepsilon^{-4} D_0^2 \exp(300000/RT)$	[59]
C-Mn	$t_{0.5srx} = 2.3 \cdot 10^{-15} \varepsilon^{-2.5} D_0^2 \exp(230000/RT)$	[85]
Ti-V	$t_{0.5srx} = 5 \cdot 10^{-18} (\varepsilon - 0.085)^{-3.5} D_0^2 \exp(280000/RT)$	[86]
Nb (<0.03%)	$t_{0.5srx} = (-5.24 + 550[Nb]) \cdot 10^{-18} D_0^2 \varepsilon^{-4.0+77[Nb]} \cdot \exp\left(\dfrac{330000}{RT}\right)$	[15]
Nb (0.045%)	$t_{0.5srx} = 1.9 \cdot 10^{-18} \dot{\varepsilon}^{-0.41} \varepsilon^{-2.8} D_0^2 \exp(324000/RT)$	[87]
Nb, Ti and Nb-Ti	$t_{0.5srx} = 9.92 \cdot 10^{-11} D_0 \varepsilon^{-5.6 D_0^{-0.15}} \dot{\varepsilon}^{-0.53} \cdot$ $\cdot \exp\left(\dfrac{180000}{RT}\right) \exp\left[\left(\dfrac{275000}{T} - 185\right)([Nb] + 0.374[Ti])\right]$	[74]

Comparing static and metadynamic kinetics (see Table 9 and Table 10), it can be seen that the latter process is highly strain rate dependent but a weak function of the deformation and temperature, whereas the opposite happens for static recrystallisation. In the *SRX* kinetics, since the driving force is the reduction of strain energy accumulated during hot working, a significant effect of the applied strain is observed. In contrast, once the steady state is achieved during dynamic recrystallisation, the balance between the softening and hardening mechanisms leads the strain energy to remain constant with strain (constant stress level). Consequently, subsequent metadynamic recrystallisation kinetics are little influenced by strain.

In conventional flat products rolling, the critical strain required to the onset of dynamic recrystallisation and the time interval between industrial passes do not favour a general occurrence of *DRX* (and therefore, metadynamic recrystallisation). Jonas has considered different scenarios where *DRX* can occur in industrial conditions [88, 89]. The first one corresponds to short interpass times, as is the case of the finishing stands in rod rolling, where dynamic recrystallisation starts once the critical strain is exceeded as a consequence of strain accumulation between passes (since there is not enough time for static recrystallisation), followed by metadynamic softening [90].

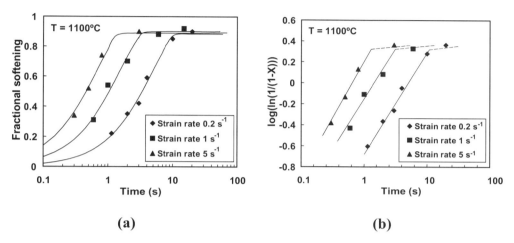

(a)   (b)

Fig. 28. a) Effect of strain rate on metadynamic recrystallisation kinetics at 1100°C in a 0.035% Nb microalloyed steel and b) corresponding Avrami plots [73].

Table 10. Equations describing time (in s) for 50% metadynamic recrystallisation.

Steel	Equation	Ref.
C-Mn	$t_{0.5mdrx} = 2.13 \cdot 10^{-6} \dot{\varepsilon}^{-0.67} \exp\left(\dfrac{133000}{RT}\right)$	[61]
Nb	$t_{0.5mdrx} = 1.05 \cdot 10^{-7} \dot{\varepsilon}^{-0.85} \exp\left(\dfrac{157000}{RT}\right)$	[87]
Nb	$t_{0.5mdrx} = 1.77 \cdot 10^{-6} \dot{\varepsilon}^{-0.62} \exp\left(\dfrac{153000}{RT}\right)$	[73]
Nb	$t_{0.5mdrx} = (12[Nb]+0.5) \cdot 10^{-4} \dot{\varepsilon}^{-0.86} \exp\left(\dfrac{92200}{RT}\right)$	[72]
Ti	$t_{0.5mdrx} = 8.9 \cdot 10^{-6} \dot{\varepsilon}^{-0.83} \exp\left(\dfrac{125000}{RT}\right)$	[91]

The second case corresponds to the strip rolling of Nb microalloyed steels. In this case, Nb in solution can retard static recrystallisation, allowing strain accumulation to take place [89]. Under certain conditions a similar behaviour will be observed in the case of thin slab direct rolling.

3.1.3. RECRYSTALLISED GRAIN SIZE

Static recrystallisation refines the austenite grain size. The statically recrystallised grain size, D_{srx}, is related to the initial grain size (it determines the density of nucleation sites) and the deformation parameters (strain, strain rate and temperature). In contrast, the grain size produced by metadynamic recrystallisation, D_{mdrx}, is only dependent on the temperature and strain rate, but not on strain. The following relationships are proposed for each range:

$$D_{srx} = D \cdot D_0^k \cdot \varepsilon^m \qquad \text{for } \varepsilon < \varepsilon_c \qquad \text{Eq. 27}$$

$$D_{mdrx} = D' \cdot Z^s \qquad \text{for } \varepsilon > \varepsilon_c \qquad \text{Eq. 28}$$

As in the previous equations, a wide range of constants has been given for the exponents in Eq. 27 and Eq. 28 depending on the steel grade. Some examples are listed in Table 11.

Table 11. Equations describing austenite grain size (in μm).

Steel	Equation	Ref.
Static		
C-Mn	$D_{srx} = 0.743 \cdot D_0^{0.67} \cdot \varepsilon^{-1}$	[92]
Nb	$D_{srx} = 1.1 \cdot D_0^{0.67} \cdot \varepsilon^{-0.67}$	[60]
Nb	$D_{srx} = 1.4 \cdot D_0^{0.56} \cdot \varepsilon^{-1}$	[93]
Metadynamic		
C-Mn	$D_{mdrx} = 2.6 \cdot 10^4 \cdot Z^{-0.23} \qquad Q = 300 \, kJ/mol$	[85]
Nb	$D_{mdrx} = 1370 \cdot Z^{-0.13} \qquad Q = 375 \, kJ/mol$	[91]
Ti	$D_{mdrx} = 39700 \cdot Z^{-0.42} \qquad Q = 372 \, kJ/mol$	[91]

3.1.4. GRAIN GROWTH AFTER RECRYSTALLISATION

At high temperatures grain growth will take place after complete recrystallisation if the interpass time during hot rolling is long enough. The reason for this growth is the reduction of the internal energy by decreasing the total austenite grain boundary area.

The evolution of the austenite grain size after recrystallisation, in isothermal conditions, is usually described by equations of the type:

$$\mathbf{D^n = D_{rex}^n + B \cdot t_q \cdot exp\left(-\frac{Q_{gg}}{RT}\right)} \qquad \text{Eq. 29}$$

where D_{rex} is the fully recrystallised grain size and t_q is the time after complete recrystallisation, normally taken as a 95% recrystallised fraction ($t_q = t_{ip} - t_{0.95srx}$, being t_{ip} the interpass time). In the equation Q_{gg} is the apparent activation energy for grain growth and n and B are constants.

Eq. 29 denotes that grain growth can take place if the interpass time is longer than the time required completing recrystallisation. However the time and the temperature are not the only parameters affecting the grain growth, the size of the recrystallised grains is also an important variable: large grains have a lower tendency to grow than finer ones. Consequently, the kinetics of grain growth will be strongly dependent on the recrystallised grain size determined by Eq. 27 and Eq. 28.

Concerning the coefficients of Eq. 29, it is expected theoretically that $n = 2$, but in practice significantly higher values have been reported. Some of them are listed in Table 12. These high values for n and the exponential dependence of grain size on temperature suggest that grain sizes below a certain limit cannot be maintained, even for very short times [94].

Taking into account the constants proposed by Hodgson et al. [15, 85] the evolution of grain size with time at a temperature of 1000°C has been drawn in Fig. 29 for the case of a plain C-Mn steel and a Nb microalloyed steel. For small initial grain sizes it is worth emphasising the solute drag effect of Nb which delays grain growth. As initial grain size increases, grain growth is smaller.

Table 12. Coefficients of Eq. 29 describing the grain growth (grain size in μm and time in seconds).

Steel	n	B	Q_{gg} (kJ/mol)	Ref.
C-Mn (T >1273K)	10	3.87×10^{32}	400	[95]
C-Mn (T < 1273K)	10	5.02×10^{53}	914	[95]
C-Mn (t < 1 s)	2	4.0×10^{7}	113	[85]
C-Mn (t > 1 s)	7	1.5×10^{27}	400	[85]
C-Mn-Nb	4.5	4.1×10^{23}	435	[15]
C-Mn-Ti	10	2.6×10^{28}	437	[15]

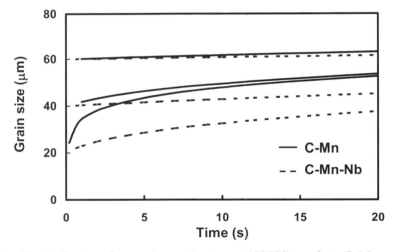

Fig. 29. Evolution of austenite grain size at 1000°C in plain C-Mn and Nb microalloyed steels considering the equations proposed in Ref. [15] and [85].

3.2. CONVENTIONAL CONTROLLED ROLLING

Microalloying has been one of the key factors in obtaining steel grades with excellent strength-toughness combinations. The good toughness is a result of the significant grain size refinement which also leads to an important strengthening. Similarly, this grain refinement can counterbalance possible loss of toughness associated with other strengthening mechanisms (such as precipitation hardening, for example). The achievement of these fine microstructures at room temperature requires a process of austenite conditioning during hot rolling prior to transformation. Controlled rolling is the procedure to implement this austenite conditioning.

In this section the main peculiarities of conventional controlled rolling will be analysed, while the characteristics of recrystallisation controlled rolling will be described in section 3.3. Both procedures are nowadays the main routes for the implementation of austenite conditioning in industrial practise.

3.2.1. AUSTENITE CONDITIONING

One of the main objectives of conventional controlled rolling is to produce a pancaked austenite microstructure. Pancaking is the result of strain accumulation in the austenite when the steel is deformed at temperatures below the non recrystallisation temperature (T_{nr}) [96, 97] (see Fig. 30). In that case, the ferrite grain size obtained after the γ-α transformation is significantly refined, compared to the one obtained from a recrystallised austenite microstructure when deformation is applied at temperatures above T_{nr} [98]. The origin of the refinement is the increase in the potential ferrite nucleation sites as a consequence of the larger specific surface, S_v, and defect density present in the deformed austenite. This benefit initially developed for ferritic steels has been successfully extended to more complex microstructures as dual phase and TRIP (transformation induced plasticity) steels.

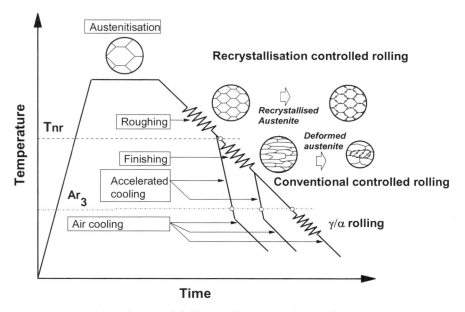

Fig. 30. Scheme of different thermomechanical processes.

The determination of the non recrystallisation temperature (T_{nr}) is a crucial step in designing controlled rolling schedules, because it defines the temperature below which strain is accumulated in the austenite (pancaking). This temperature is a result of the interaction between deformation, recrystallisation and precipitation. The following two mechanisms have been proposed as being responsible: the solute drag effect due to solute atoms and the pinning effect due to strain induced precipitates, the latter usually exerting the stronger effect [99 - 101]. In order to obtain a maximum refinement in the final ferrite microstructure, once the roughing is completed after the first rolling stands, the final rolling pass parameters must be optimised to achieve as much pancaking as possible in the microstructure.

When austenite containing microalloying elements in solution is thermomechanically worked at falling temperature, the degree of supersaturation may become high enough for dislocations, generated during deformation, to serve as sites for the precipitation of carbonitrides before recrystallisation reduces the dislocation density to an innocuous level. Fine precipitation then inhibits further recrystallisation during the remainder of the processing.

In this context, the first objective is to consider the microalloying elements which can precipitate during hot working. In Fig. 31 the solubility product of different nitrides and carbides are drawn [102 - 106]. Similarly, taking into account some typical examples of steel grades, the evolution of the precipitated percentage of a microalloyed element as a function of temperature, in equilibrium conditions, is illustrated in Fig. 32. From the figure it is obvious that the element most likely to precipitate during rolling is Nb. In contrast, Ti is in most cases precipitated before rolling, while vanadium precipitation will typically occur after rolling has finished.

This special behaviour of Nb is the reason why this element is thought to be the most convenient to achieve austenite conditioning. Nevertheless, the presence of Nb in the steel chemical composition does not guarantee that the required austenite pancaking is achieved by itself, as it is necessary to adequately define the different process variables involved in hot rolling.

Fig. 31 and Fig. 32 show that Nb can precipitate during hot rolling, but it must be pointed out that in both figures kinetics are not included. It is worth emphasising the fact that in undeformed austenite Nb(C,N) precipitation is relatively slow.

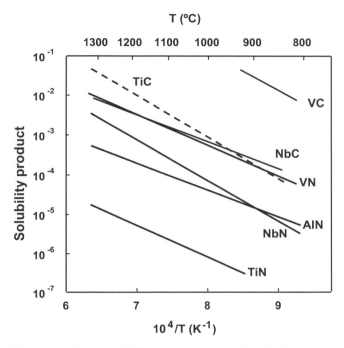

Fig. 31. Solubility products of different nitrides and carbides in austenite [102 - 106].

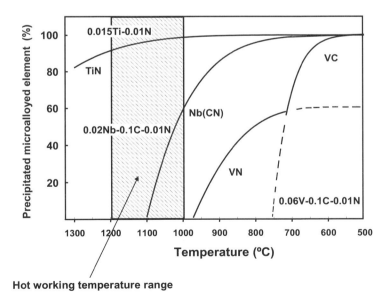

Fig. 32. Examples of precipitated elements as a function of temperature, in equilibrium conditions, for three microalloyed steels containing Ti, Nb and V.

To take into account the effect of strain on subsequent precipitation kinetics, Dutta and Sellars proposed the following expression to determine the time required for 5% precipitation [111] (5% relative to the equilibrium precipitation at the corresponding temperature):

$$t_{0.05p} = A \cdot [C]^{-1} \cdot \varepsilon^{-1} \cdot Z^{-0.5} \cdot \exp\left(\frac{Q_d}{RT}\right) \cdot \exp\left(\frac{B}{T^3 \cdot [\ln k_s]^2}\right) \qquad \text{Eq. 30}$$

where [C] represents the amount of solute available for precipitation, Q_d is the activation energy for diffusion of the rate controlling solute species in the matrix, and k_s is the supersaturation ratio at the deformation temperature of the austenite, defined as:

$$k_s = \frac{[Nb]\left[C + \frac{12}{14}N\right]_{sol}}{10^{\left(2.26 - \frac{6770}{T}\right)}} \qquad \text{Eq. 31}$$

Eq. 30 was fitted to experimental data for Nb-bearing steels, and the values of $A = 3.10^6$ and $B = 2.5.10^{10}$ were found to give the best fit for reheated austenite. Based on the same equation, Abad et al. [93] determined the constants by fitting experimental results including microstructural conditions similar to those found in thin slab casting and direct charging processes (coarse grain size and high supersaturation levels), leading to the following expression:

$$t_{0.05p}(s) = 5.3 \times 10^{-7} [Nb]^{-1} \varepsilon^{-1} Z^{-0.5} \exp\left(\frac{270000}{RT}\right) \exp\left(\frac{1.3 \times 10^{10}}{T^3 (\ln k_s)^2}\right),$$

$$\text{Eq. 32}$$

$$\text{with } Z = \dot{\varepsilon} \exp\left(\frac{341000}{RT}\right)$$

Strain induced precipitation kinetics defined by the previous equations significantly change when compared to undeformed austenite conditions, as can be observed in Fig. 33. Similarly, the time required for precipitation to start is lower as larger strains are applied and/or Nb content increases (see Fig. 34).

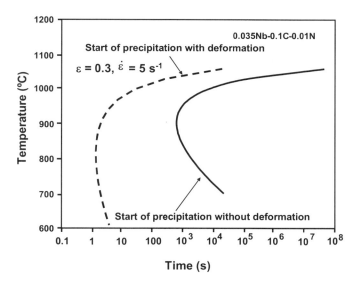

Fig. 33. Change in Nb(C,N) precipitation kinetics with applied strain.

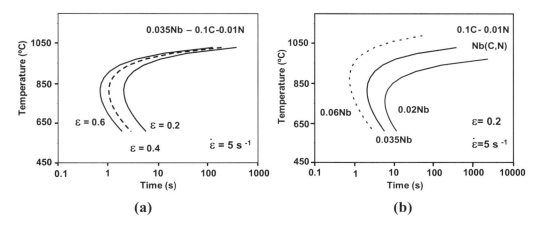

Fig. 34. Influence of applied strain and Nb content on Nb(C,N) precipitation start kinetics (in agreement with Eq. 32).

3.2.2. NON RECRYSTALLISATION TEMPERATURE

The interaction between precipitation and recrystallisation can be considered in terms of the driving forces related to both mechanisms [101, 107, 108]. The driving force for recrystallisation is the stored energy of deformation and it can be estimated from the equation [108]:

$$\mathbf{F_{rex}} = \mathbf{12.5}(\Delta\sigma)^2\mu^{-1} \qquad\qquad\qquad\qquad\qquad \text{Eq. 33}$$

where μ is the shear modulus (it is temperature dependent [109]) and $\Delta\sigma$ the increase in flow stress during work hardening.

Different models [108] have been proposed for the estimation of the precipitation pinning force F_{pin}, depending on the assumptions made about which particles interact with the motion of the boundary, as shown in Table 13. From several analyses carried out with different steel grades, it seems that the flexible boundary model appears to be the most convenient to characterise F_{pin} [77, 110].

When the pinning force exerted by the precipitates overcomes the stored energy of deformation (see Fig. 35), $F_{pin} > F_{rex}$, recrystallisation should stop completely [107]. This is the case when strain induced precipitation takes place at the same time as recrystallisation.

Table 13. Precipitation pinning force equations.

Model	F_{pin} (MPa)
Rigid boundary	$\dfrac{6\gamma f_v}{\pi r}$
Flexible boundary	$\dfrac{3\gamma f_v^{2/3}}{\pi r}$
Subgrain boundary	$\dfrac{3\gamma f_v l}{2\pi r^2}$
f_v: precipitate volume fraction; γ: the interfacial energy per unit area of boundary (0.8 J/m^2); r: particle radius; l: average subgrain boundary intercept distance.	

Dutta and Sellars [111] modelled the interaction between recrystallisation and strain induced precipitation to predict the temperature at which, for a given strain, precipitation occurs sufficiently quickly to stop recrystallisation. This temperature is known as the recrystallisation stop temperature (*RST*). The criterion they used to define this temperature was that 5% precipitation should be reached before 5% recrystallisation occurred. This procedure is followed in

Fig. 36 to define the *RST* temperature for the case of a 0.035% Nb microalloyed steel.

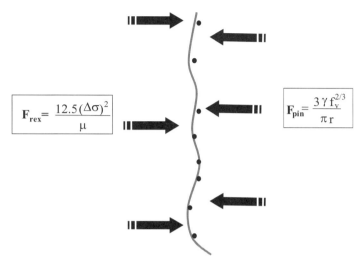

$$F_{rex} = \frac{12.5(\Delta\sigma)^2}{\mu}$$

$$F_{pin} = \frac{3\,\gamma\,f_v^{2/3}}{\pi\,r}$$

Fig. 35. Scheme showing the interaction between the driving force for recrystallisation and the pinning force exerted by precipitates.

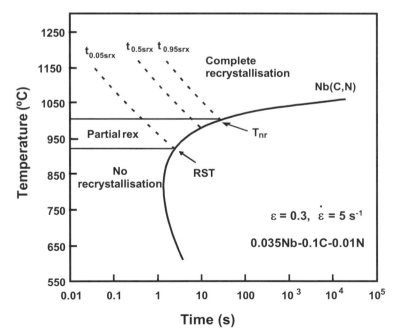

Fig. 36. Definitions of recrystallisation stop temperature (RST) and non recrystallisation temperature (T$_{nr}$).

Similarly, it is possible to determine the non recrystallisation temperature (T_{nr}), which is defined as the temperature above which complete recrystallisation takes place. The criterion usually adopted to determine this temperature is that 95% recrystallisation is reached before 5% precipitation occurs (see Fig. 36). The time to achieve 5% or 95% recrystallisation can be calculated from the time required for 50% recrystallisation, given by Eq. 23 if the Avrami exponent in Eq. 22 is known.

Taking into account Eq. 23 and Eq. 32 it is possible to evaluate the variations that occur in T_{nr} temperature when different Nb contents or processing conditions are considered. For example, Fig. 37 illustrates the increase in T_{nr} temperature when Nb content changes from 0.035% to 0.06%.

Between *RST* and T_{nr} temperatures there is an interval where a mixture of microstructures of recrystallised and deformed austenite grains coexist (Fig. 36). From a practical point of view, it is convenient to make sure that the final rolling conditions do not take place in this region. Otherwise, the possibilities of some heterogeneities appearing in the final microstructure after transformation are significantly higher.

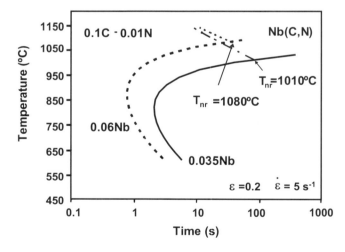

Fig. 37. Variation in T_{nr} temperature when increases Nb content from 0.035% to 0.06%.

Jonas and co-workers developed a method [112, 113] to determine the T_{nr} based on multipass torsion tests performed under continuous cooling conditions. Fig. 38 shows the typical stress-strain curves obtained in a multipass torsion test carried out with a Nb microalloyed steel. This kind of test is widely used to simulate industrial hot rolling processes. Under these conditions, the T_{nr} is a function of process variables which include pass-strain and strain rate, pass temperature and interpass time, with all of them having a great effect on microstructural evolution during the rolling process.

An example of a multipass torsion test stress-strain curve is shown in Fig. 38. From the figure, the mean flow stress (*MFS*) corresponding to each pass of the torsion test was calculated by numerical integration and plotted against the inverse absolute temperature in Fig. 39. The *MFS* has been defined as the area under each stress-strain curve divided by the strain difference. Three different regions can be clearly distinguished in both, Fig. 38 and Fig. 39:

1) Region I: where it is supposed that complete recrystallisation (100%) between passes takes place and the stress increases from pass to pass are only due to the drop in temperature.

2) Region II: where recrystallisation between passes is inhibited by strain induced precipitation. The stress increases more rapidly due to both, the drop in temperature and the accumulation of strain.

3) Region III: corresponds to the austenite-ferrite region. The stress reduction results from the start of the transformation of austenite to softer ferrite phase (A_{r3} temperature).

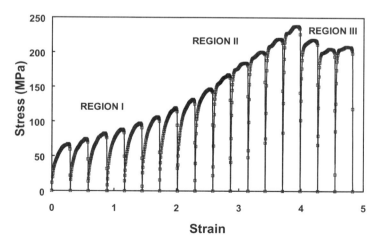

Fig. 38. Stress strain curves obtained in a 17 pass torsion test corresponding to a 0.035% Nb microalloyed steel (deformation conditions: T_{soak} =1200°C (15 min), ε=0.3, t_{ip}=30 s) [93].

Following the method developed by Jonas and co-workers, the value of T_{nr} can be determined from the intersection between the regression lines of the points corresponding to regions I and II.

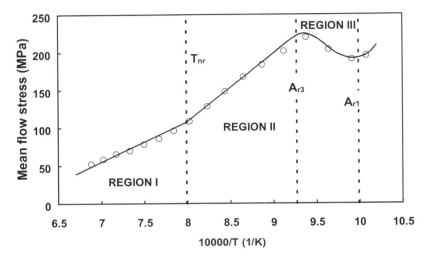

Fig. 39. Mean flow stress against inverse absolute temperature obtained with the data of Fig. 38 [93].

As previously mentioned, the non recrystallisation temperature depends on several parameters, such as pass-strain and interpass time. Referring the interpass time, three different regions can be distinguished [112], as illustrated in the example of Fig. 40. For very short times, precipitation is unable to occur and solute drag is the only mechanism delaying recrystallisation. In this condition the T_{nr} decreases with increasing interpass time. At intermediate interpass times (second region), the occurrence of strain induced precipitation retards recrystallisation and leads to an increase of the T_{nr}, that remains nearly constant in this range of interpass times. Finally, a further increase in the time means that precipitate coarsening is able to occur, weakening retardation of recrystallisation (see Table 13), and the T_{nr} decreases again (third region).

Similarly, the microalloying content significantly affects the value of T_{nr}, this influence becoming more complex in the case of multiple microalloying additions. For example, in Fig. 41 the influence of different Ti contents in Ti-Nb steels is compared with the results obtained with Nb steels [115]. For similar contents of Nb, the T_{nr} of the Nb-Ti steels is significantly lower than that observed for Nb steels at all strains. This behaviour could be related to several factors contributing to the decrease in the supersaturation level when Ti is present in addition to Nb [114].

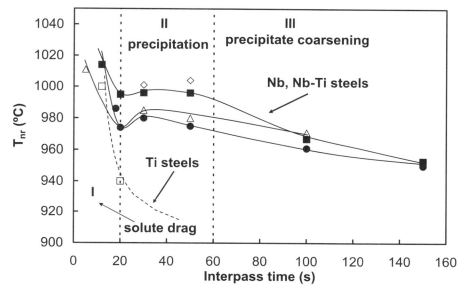

Fig. 40. The three different ranges of interpass time effect on the non recrystallisation temperature in Nb, Ti and Nb-Ti microalloyed steels [115].

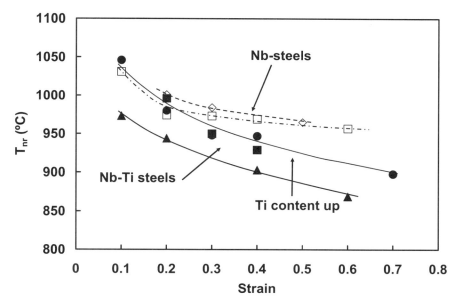

Fig. 41. Influence of T_{nr} on the pass strain in Nb and Nb-Ti steels [115].

3.2.3. PARTIALLY RECRYSTALLISED MICROSTRUCTURE

In multipass rolling, it is possible to have partial recrystallisation after a pass strain ε_1, if the time between passes is not long enough to complete recrystallisation. This introduces a mixed microstructure before the next deformation pass, ε_2. To cope with this aspect the "uniform softening method" has been proposed [116]. This method considers a single average grain size with an effective strain as follows:

$$\varepsilon_{eff} = \varepsilon_2 + \lambda(1 - X)\varepsilon_1$$

Eq. 34

given by the strain of the second pass (ε_2) plus the residual strain from the first pass ($\lambda(1-X)\varepsilon_1$), λ being a constant which is taken as 0.5 for Nb steels [117] and X is the recrystallised fraction between passes 1 and 2.

In this method the partially recrystallised microstructure is described by an average grain size. Different expressions have been proposed to calculate this grain size [116, 118]. Some authors calculate a mean grain size by the application of the law of mixtures, taking into account the fact that the microstructure is a mixture of recrystallised and unrecrystallised grains [63, 119]. Other authors utilise an average grain size equal to the fully recrystallised grain size which would be obtained in the case of complete recrystallisation [117, 120].

Independently of the softening mechanism involved (static or metadynamic), partially recrystallised microstructures are characterized by the recrystallising grain size, D_r. Theory predicts that if site saturation holds, there are no shape changes, the distribution of recrystallising grains remains stable during recrystallisation and no grain coarsening takes place, the evolution of the mean recrystallising grain size D_r with time can be fairly described by the following relationship [121]:

$$D_r = D_{rex} \cdot X^{1/3}$$

Eq. 35

where D_{rex} represents the final recrystallised grain size calculated for the corresponding post-dynamic softening mechanism, static (D_{srx}) or metadynamic (D_{mdrx}) recrystallisation. Nevertheless, it has been reported that from a practical point of view Eq. 35 results are appropriate to describe the evolution of D_r in a simple manner, even if theoretical conditions do not hold exactly [121, 122].

The effective size of unrecrystallised grains, D_u, can be described with the help of the following expression proposed by Anelli [123]:

$$\mathbf{D_u} = 1.06 \exp(-\varepsilon)(1-X)^{1/3}\,\mathbf{D_0}$$

Eq. 36

where ε is the applied strain in each pass and X is the total recrystallised volume fraction: static (X_{srx}), metadynamic (X_{mdrx}) or the sum of both.

Recently Fernandez et al. [122] have observed that the uniform softening method, assuming an "average" microstructure with a mean grain size equal to that of the fully recrystallised material ($\overline{D} = D_{rex}$), leads to reasonable results when the recrystallised fractions are larger than $X \approx 0.4$. However, the accuracy of this method for coarse austenite microstructures, in terms of recrystallisation time predictions, may depend to a large extend on the deformation schedule. Large deformations increase the difference in size between recrystallised and unrecrystallised grains. Under these conditions significant differences may be produced between the recrystallisation time calculated for an "average" grain size and for the unrecrystallised grain size.

3.3. RECRYSTALLISATION CONTROLLED ROLLING. EFFECT OF TITANIUM

Steels with small additions of Ti exhibit limited grain growth, even at high temperatures. Finely dispersed TiN particles can exert a pinning effect on the grain boundaries inhibiting their migration. Fig. 42 shows the evolution of the austenite grain size as a function of the reheating temperature for different microalloyed steels [124]. In the figure the beneficial effect of Ti addition is clearly seen. Nevertheless, to obtain a good overall grain control a minimum volume fraction of fine TiN particles is required. To reach the maximum yield of fine TiN precipitates, the formation of coarse particles during solidification (> 0.2 μm) should be avoided, taking into account the fact that these particles do not contribute to grain control and moreover, may even be prejudicial for toughness. In this sense, several practical recommendations need to be considered: a low amount of Ti with hypostoichiometric Ti/N relationships (Ti/N<3.4), rapid solidification rates and post-solidification cooling rates (continuous casting is recommended) [125]. If these conditions are not fulfilled, only a small fraction of Ti may stay available to perform an efficient microstructural control (see Fig. 43).

"Recrystallisation controlled rolling" is based on the application of rolling passes above the non recrystallisation temperature, T_{nr}, and involves the control of austenite grain growth during reheating, rolling and after rolling. The TiN particles are very suitable for the prevention of austenite grain growth both during reheating and rolling. Consequently making use of these advantages it is possible to achieve, with the help of repeated recrystallisations, a fine recrystallised austenite grain before transformation. Fig. 44 shows examples of the evolution of the austenite grain size in a plain C-Mn steel and in a Ti microalloyed steel.

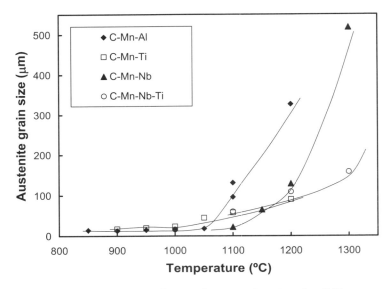

Fig. 42. Austenite grain size evolution during reheating for different steels [124].

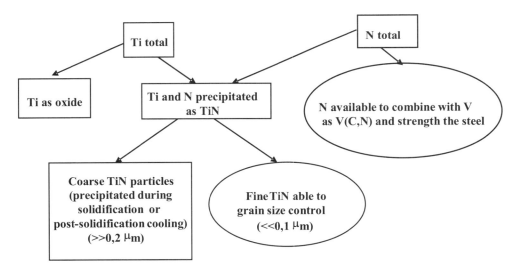

Fig. 43. Different possible combinations of Ti microaddition with oxygen, nitrogen and vanadium.

Because grain growth between passes is completely avoided at the exit of the last rolling pass a significant grain refinement is achieved in the Ti steel. Following this procedure, it is possible to obtain relatively fine ferrite grain sizes with relatively high finishing rolling temperatures (900-1050°C) [18]. This implies that this route is a good alternative for those rolling mills where the capacity to perform conventional controlled rolling is limited.

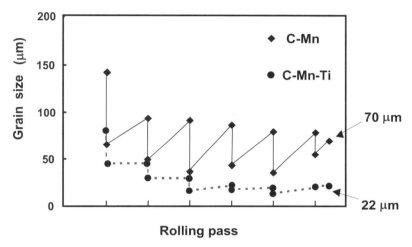

Fig. 44. Schematic illustration of austenite grain size evolution during hot rolling comparing a plain C-Mn steel and a Ti microalloyed steel.

Recrystallisation controlled rolling is commonly applied to Ti-V microalloyed steels. While the microaddition of Ti is related to grain growth control, microalloying with vanadium provides a fine V(C,N) precipitation after hot rolling that significantly contributes to the strength of the steel. As a result of an adequate combination of both mechanisms (grain refinement and precipitation strengthening), as-hot rolled microstructures with high strength and good toughness properties ($\sigma_y \sim 500$ MPa and ITT $< -60°C$) are produced [126].

3.4. MEAN FLOW STRESS

The mean flow stress during hot rolling is a relevant factor that needs also to be considered. It provides information about the loads required during each rolling pass but it may also be a useful tool for the identification of the softening/hardening mechanisms that intervene in an industrial rolling mill [70].

Several expressions have been proposed to calculate the mean flow stress as a function of composition and deformation parameters [127, 128]. One of these is the Misaka equation (Eq. 37), often employed to determine the MFS of C-Mn steels during hot strip rolling [127]:

$$\text{MFS(MPa)} = 9.8 \exp\left(0.126 - 1.75[C] + 0.594[C]^2 + \frac{2851 + 2968[C] - 1120[C]^2}{T}\right) \cdot$$
$$\cdot \varepsilon^{0.21} \dot{\varepsilon}^{0.13} \qquad \text{Eq. 37}$$

where *[C]* is the carbon content and *T* is the temperature in K. Based on this equation a new modified Misaka equation that takes into account the effect of different alloying elements, such as Mn, Nb and Ti, was developed by Minami et al. [67]. This relationship is displayed in Eq. 38. It introduces strengthening factors for the different alloying elements present in solution in austenite.

$$\text{MFS(MPa)} = 9.8 \exp\left(0.126 - 1.75[C] + 0.594[C]^2 + \frac{2851 + 2968[C] - 1120[C]^2}{T}\right) \cdot$$
$$\cdot (0.768 + 0.51[Nb] + 0.137[Mn] + 4.127[Ti]) \varepsilon^{0.21} \dot{\varepsilon}^{0.13} \qquad \text{Eq. 38}$$

MFS depends also on deformation parameters, i.e. the amount of strain and strain-rate, both variables included in Eq. 38.

On the other hand, from the data of the industrial mill it is possible to calculate the MFS according to the Sims method as follows:

$$\text{MFS}_{\text{Sims}} = P \left/ \left(\frac{2}{\sqrt{3}} w(R(H-h))^{1/2} \cdot Q \right) \right. \qquad \text{Eq. 39}$$

where *P* is the roll force, *w* the plate width, *R* the radius of the roll, *H* initial thickness of the plate, *h* final thickness and *Q* a multiplier factor dependent on the geometry of the rolling pass [129]. This equation can be corrected to include the effect of flattened roll radius and the redundant strain with the help of different expressions defined in Ref. [130].

In industrial conditions the strains and strain rates can change from one stand to the next. In order to compare *MFS* values corresponding to different stands in this situation it is necessary to correct them (*MFS$_{corr}$*) to constant values of strain and strain rate [67]. Based on strain and strain rate exponents shown in Eq. 38, the following equation can be used [67]:

$$\text{MFS}_{\text{corr}} = \text{MFS}_{\text{Sims}} \cdot (\varepsilon_{\text{ref}} / \varepsilon_{\text{pass}})^{0.21} \cdot (\dot{\varepsilon}_{\text{ref}} / \dot{\varepsilon}_{\text{pass}})^{0.13} \qquad \text{Eq. 40}$$

where ε_{ref} and $\dot{\varepsilon}_{ref}$ are the reference values selected.

Comparing the *MFS* values obtained from the mill data and those calculated from Eq. 38, it is possible to determine whether complete static recrystallisation, accumulation of strain or maybe dynamic recrystallisation are occurring in industrial conditions. The different situations that can occur are schematised in Fig. 45.

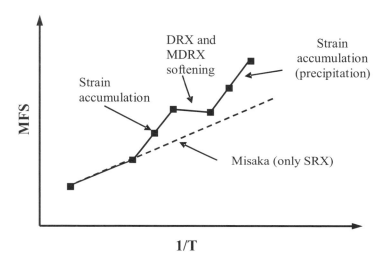

Fig. 45. Scheme showing the evolution of MFS as a function of inverse temperature.

If recrystallisation is complete between passes, the *MFS* obtained with the mill data will be properly predicted by the corrected Misaka equation. Once strain accumulation occurs, the slope will increase significantly deviating from the Misaka predictions. This increase in the MFS can also be taken into account in the Misaka equation by considering the accumulated strain instead of the nominal value. If this strain accumulation is due to very short interpass times (combined with some additional effect, such as solute drag for example), once the ε_c critical strain is reached, dynamic recrystallisation followed by metadynamic softening can occur, leading to a decrease in *MFS*. Finally, if strain induced precipitation stops recrystallisation, again a significant increase in the *MFS* will be observed, as shown in the scheme of Fig. 45.

3.5. PHASE TRANSFORMATION DURING COOLING. PRECIPITATION HARDENING

After deformation the softening and hardening effects discussed in the previous sections will determine the condition of the austenite. During cooling this austenite will transform into lower temperature phases. The factors that affect the resulting microstructure are those determined by the deformation history: the final austenite grain size, the retained strain and the external variables: composition and cooling schedule. Depending on the combination of these factors the proportion of the different phases (ferrite, pearlite, acicular phases...) as well as the ferrite grain size will change. With appropriately selected parameters a wide range of final microstructures can be obtained as shown in the scheme of Fig. 46: mainly ferritic microstructures (HSLA steels), dual phase and multiphase steels (TRIP).

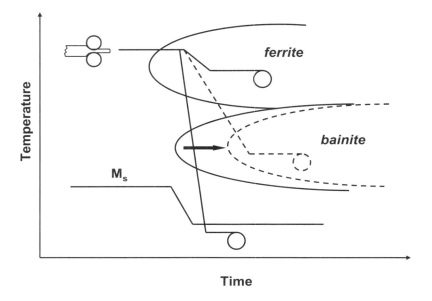

Fig. 46. Scheme showing different cooling and coiling strategies to achieve different final microstructures (HSLA, dual phase and TRIP steels).

In the following, the main aspects related to austenite/ferrite transformation will be considered.

3.5.1. AUSTENITE TRANSFORMATION

Concerning the influence of microstructure prior to transformation in the case of recrystallised austenite, ferrite nucleates mainly at the grain corners, edges and surfaces (see scheme of Fig. 47), in that order of preference. The final steel microstructure will be refined by reducing the austenite grain size prior to transformation, thus providing a higher density of nucleation sites. An example of the transformation evolution is shown in Fig. 48 for a 0.05% Nb microalloyed steel [131]. Similarly, increasing cooling rate delays the transformation to lower temperatures, increasing the density of sites where the nucleation occurs.

If the austenite has a pancaked form with some degree of accumulated strain, there will be a higher density of sites available for ferrite nucleation. In addition to the grain boundaries, the defects generated inside the austenite grains, such as deformation bands and twin boundaries, also provide nucleation sites as is shown schematically in Fig. 47. The effect of accumulating strain in the nucleation site density and in the ferrite grain refinement is clearly apparent in the example shown in Fig. 49 [131]. Starting with a similar initial austenite grain size as in the case of Fig. 48, the pancaked microstructure results in a refinement of the ferrite grain size.

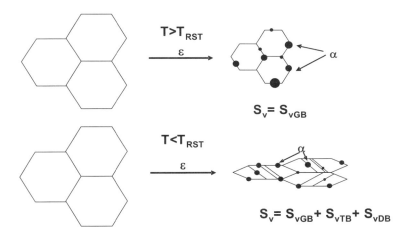

Fig. 47. Scheme showing the nucleation sites in recrystallised and in deformed austenite microstructures.

Some authors define an effective grain boundary area per unit volume, S_v, where besides the grain boundaries the contribution of intragranular defects present in the austenite is also included [132]:

$$S_V = S_{VGB} + S_{VTB} + S_{VDB} \qquad \text{Eq. 41}$$

where the subscripts GB, TB and DB denote the contribution to the total S_v from the grain boundaries, twin boundaries and deformation bands. In the case of recrystallised grains only the first term becomes significant (Fig. 47) and for a given D grain size: $S_v = 2/D$. In the case of deformed grains different expressions can be found in the literature as a function of the deformation mode. In the case of flat rolling (plain strain conditions), S_v can be determined from the following equation:

$$S_v = \frac{1}{D_\gamma}\left(0.429\exp\left(-\frac{\sqrt{3}}{2}\varepsilon_{acc}\right)+0.571+\exp\left(\frac{\sqrt{3}}{2}\varepsilon_{acc}\right)\right)$$ 　　　　Eq. 42

where D_γ is the austenite grain size prior to deformation and ε_{acc} represents the amount of strain applied.

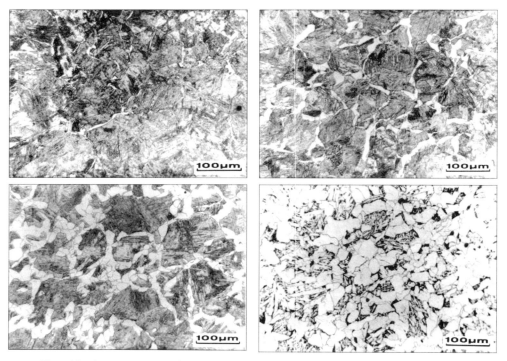

Fig. 48. Austenite transformation in a 0.05% Nb microalloyed steel with a recrystallised austenite grain size of 43 μm (cooling rate 1°C/s) [131].

There are different published data that confirm that for a given S_v value the transformation from an elongated austenite microstructure gives even finer ferrite grain sizes than from an equiaxed one [98] (see example in Fig. 50 [133]). This means that in the case of deformed austenite there must be additional nucleation sites available. Nevertheless, some authors have observed that these differences seem to decrease for the case of high S_v values in vanadium bearing steels [134].

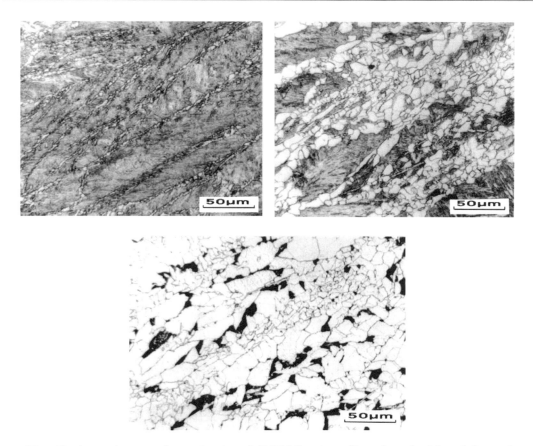

Fig. 49. Austenite transformation in a 0.05% Nb microalloyed steel with a deformed austenite (D_γ = 40 μm and accumulated strain ε_{acc} = 1; cooling rate 1°C/s) [131].

From the point of view of mechanical behaviour, the ferrite grain size prediction becomes a very important issue. There are different empirical expressions that are commonly used to quantify the mean ferrite grain size ($D_{\alpha o}$) as a function of the prior austenite grain size and cooling rate. One of the most widely accepted is the equation proposed by Beynon and Sellars [135]:

$$D_{\alpha o} = a + b \dot{T}^{-0.5} + c\left(1 - \exp\left(-0.015 D_\gamma\right)\right)$$

Eq. 43

where both grain sizes are in μm, $\dot{T}$ represents the cooling rate under continuous cooling conditions and a, b and c are constants depending on the steel chemical composition. The values of the constants for different steel families are listed in Table 14.

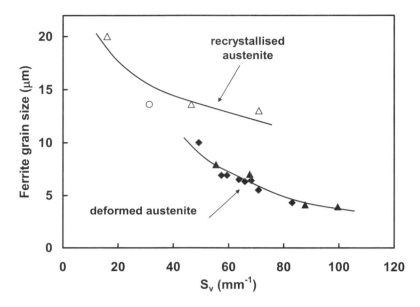

Fig. 50. Influence of S_v on ferrite grain size depending on the initial austenite microstructure in a Nb microalloyed steel [133].

The previous equation was modified by Hodgson and Gibbs [15] to incorporate the effect of carbon and manganese as follows:

$$D_{\alpha o} = \left(-0.4 + 6.4C_{eq}\right) + \left(24.2 - 59C_{eq}\right)\dot{T}^{-0.5} + 22\left(1 - \exp\left(-0.015D_{\gamma}\right)\right)$$ Eq. 44

where $C_{eq} = C + Mn/6$.

With the help of Eq. 43, the beneficial effect of accelerated cooling during transformation is illustrated in Fig. 51 for the case of a plain C-Mn steel.

Both Eq. 43 and Eq. 44 consider that the initial austenite microstructure is completely recrystallised. In order to quantify the effect of any retained strain ε_{acc} present in the austenite, a separated term is included, as follows:

$$D_{\alpha} = D_{\alpha o}\left(1 - d\,\varepsilon_{acc}^{e}\right)$$ Eq. 45

Here again d and e are constants whose values are listed in Table 14.

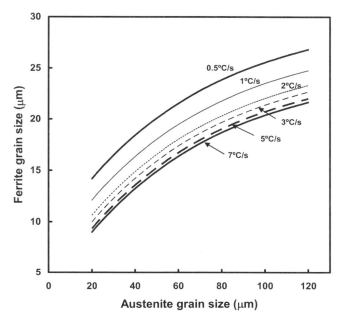

Fig. 51. Influence of continuous cooling rate in ferrite grain refinement in plain C-Mn steels considering Eq. 43.

Table 14. Values of constants in Eq. 43 and Eq. 45.

Steel	a	b	c	d	e	Ref.
C-Mn	1.4	5	22	0.45	0.5	135
Nb	2.5	3	20	0.45	0.5	135
Nb	4.5	3	13.4	0.50	0.47	98
Nb	29	-5	20	0.8	0.15	136
Ti-V	3	1.4	17	0.45	0.5	135

In the case of Nb microalloyed steels, the equation corresponding to ref. [98] provides very good predictions with different experimental results in both recrystallised and non-recrystallised conditions [137], while the expression in ref. [136] fits better when some degree of retained strain is present.

Finally, there are two more aspects related to ferritic microstructures which need to be considered. First, at high coiling temperatures (between 650 and 800°C) coarser ferrite grain sizes than those predicted by previous equations are often observed. It has been suggested that this could be a consequence of the incomplete transformation of ferrite before coiling [138]. In order to predict the ferrite grain size under these conditions, it has first been proposed to determine the non transformed austenite fraction during continuous cooling and then assume that it will transform isothermically during coiling. The mean grain size is determined by applying the law of mixtures between freshly and previously transformed ferrite grains.

Secondly, mesotexture needs to be taken into account, mainly when toughness is an important requirement. In this context it should be noticed that when an analysis of misorientations between adjacent ferrite grains is made (for example by EBSD-OIM technique) it is frequently found that the grain size determined by high angle boundaries (>15°) is significantly coarser than that quantified by optical microscopy. It must be remembered that, as was previously mentioned in Chapter 2, in general, a minimum crystallographic misorientation of 12-15 degrees is assumed as appropriate to define the microstructural unit that controls cleavage propagation. An example of a ferritic microstructure differentiating low and high angle boundaries by EBSD technique is shown in Fig. 52 [139]. As a consequence, when high toughness requirements are needed, the predictions obtained by previously considered equations should be completed by a mesotexture analysis.

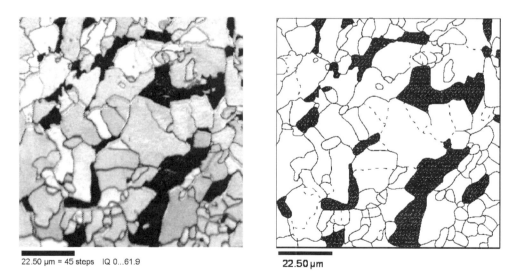

22.50 µm = 45 steps   IQ 0...61.9   22.50 µm

Fig. 52. Example of differences between low (dotted lines) and high (full lines) angle boundary misorientations in a ferritic microstructure (0.05% Nb- 0.08% V microalloyed steel, $D_\gamma = 40$ µm and accumulated strain $\varepsilon_{acc} = 1$) [139].

3.5.2. PRECIPITATION

Another aspect that needs to be considered is the precipitation strengthening when Nb and V are added. In the first case, solute Nb, which has not been precipitated during austenite processing, may form niobium carbides. In the second case, because of its solubility product, vanadium will remain mainly in solution before transformation, thus able to precipitate in the form of V(C,N) particles afterwards. It is worth emphasising that precipitation strengthening is the main reason for vanadium microalloying in a majority of plate and strip steel applications.

Precipitation can occur in two different moments during transformation. The first case corresponds to interphase precipitation and it forms repeatedly in the moving γ/α boundary as the transformation front moves through the austenite. These precipitates are arranged in sheets parallel to the instantaneous position of the γ/α interface. At lower temperatures, a general randomly distributed precipitation takes place behind the migrating γ/α boundary in the supersaturated ferrite. Examples of both types of precipitation are shown in Fig. 53 [140].

In the case of niobium, precipitation during transformation and its quantitative effect on strengthening remains an open subject [141]. Niobium carbide precipitation, during or after transformation, is dependent on the finishing rolling temperature and cooling conditions. From the analysis done in ref. [141] it can be concluded that:

- Finishing rolling temperatures of about 900°C with coiling temperatures above 680°C and coil cooling at about 0.5°C/min are conditions where precipitation has been observed. If isothermal coiling conditions are considered, at temperatures equal to or above 600°C, precipitation occurs (see example in Fig. 54).

- In contrast, processing cooling rates of more than 10°C/s from the finishing rolling temperature and coiling temperatures in the interval between 550 and 650°C (with coil cooling rates of 0.5°C/min or higher) do not show ferrite precipitation.

Referring vanadium precipitation, it is understood that an increase in nitrogen content contributes to higher strength values [18]. A higher nitrogen content leads to finer and more abundant V(C,N) particles increasing, in agreement with Eq. 5, the precipitation strengthening for a given vanadium content. This finer precipitation is due to the fact that the chemical driving force for precipitation increases as more nitrogen is dissolved in ferrite and therefore, nucleation rate enhances (see Fig. 55).

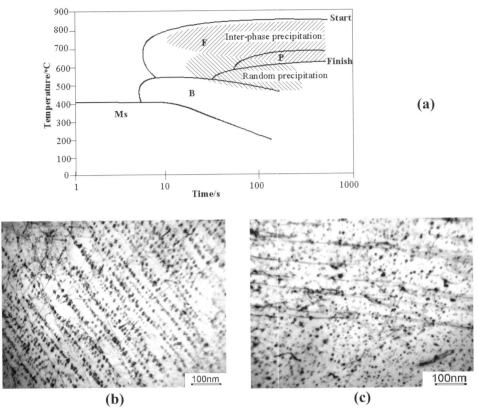

(a)

Fig. 53. a) CCT diagram indicating regions of b) interphase and c) random V(C,N) precipitation [140].

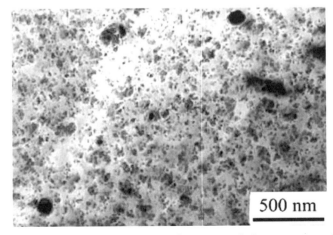

Fig. 54. Example of Nb precipitation in ferrite in a laboratory test (0.056% Nb microalloyed steel cooled at 10°C/s from final deformation temperature of 900°C and hold 1 h at 650°C) [142].

Finally, there is another aspect related to vanadium concerning its availability to refine the ferrite grain size by increasing the nucleation sites during transformation. The beneficial effect of vanadium in promoting additional ferrite nucleation has been attributed to the precipitation of VN particles both inside the austenite grains for the case of idiomorph ferrite [143, 144] and at the grain boundaries for allotriomorph ferrite [144]. This precipitation prior to transformation requires the application of specific rolling conditions. The operating mechanism seems to be related to the creation of coherent interfaces to ferrite, lowering the interfacial energy associated with ferrite nucleation and therefore promoting it [145].

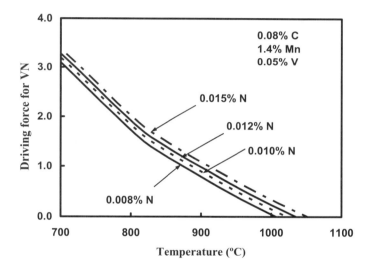

Fig. 55. Chemical driving force for precipitation of VN as a function of nitrogen content (values determined with Thermo-Calc software)

There are different works confirming that both MnS and CuS become active nucleation sites in V microalloyed steels, when V(C,N) coprecipitation takes place on the aforementioned particles [146 - 149]. One example of a MnS+V(C,N) particle as an active site for ferrite nucleation is shown in Fig. 56 [150]. Similarly, the ferrite grain size refinement that can be obtained with this complementary nucleation is shown in Fig. 57 for the case of a 0.05% V microalloyed steel. In the figure, the grain size refinement associated with additional ferrite nucleation at austenite grain boundaries (at coarse V(C,N) particles) and inside the grains (at MnS+V(C,N) complex particles) is quantified, the results being compared with those obtained with a plain C-Mn steel [150].

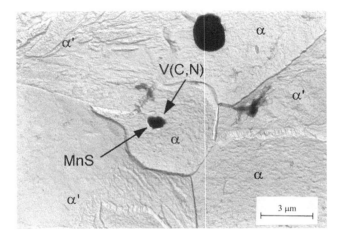

Fig. 56. TEM micrograph showing nucleation of allotriomorph ferrite associated to complex MnS+V(CN) particles [150].

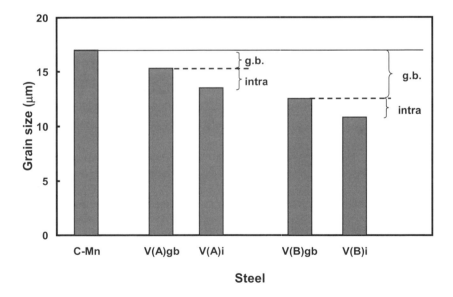

Fig. 57. Quantification of the ferrite grain size refinement promoted by vanadium enhancing allotriomorphic (gb) and intragranular (i) nucleations. The results obtained with a 0.05% V microalloyed steel with two different treating conditions are compared with the data corresponding to a plain C-Mn steel [150].

4. TSDR: CONTINUOUS CASTING AND TUNNEL FURNACE

In the previous chapter the main microstructural changes acting during hot rolling were described. Based on the considerations mentioned above, the design of adequate rolling schedules is the key factor for achieving the final mechanical properties required for each specific application. Before evaluating how these microstructural changes happen in the case of thin slab direct rolling, some of the peculiarities associated to this process and that take place before rolling are analysed in this chapter.

4.1. CONTINUOUS CASTING

One of the main peculiarities of CSP process is the mould itself. The mould has a funnel-shaped section across its upper central part which provides the possibility of properly introducing the submerged entry nozzle. This funnel section gradually narrows down into a rectangular cross section which gives the final shape of the thin slab (other main mould types used in thin slab technologies are the parallel mould and the lens-shaped mould [151]). The casting speed is typically about 4-6 m/min, 2-4 times faster than during conventional casting (between 0.8-1.8 m/min) [152]. Once the thin slab exits the mould, water-spray and air-mist nozzles act immediately in order to complete solidification.

There are two main aspects that need to be evaluated in the thin slab continuous casting as they can affect final properties:

– Possibility of smaller microsegregations compared to thick slabs.

– Risks of transverse cracking.

In both cases microalloying additions can exert an important effect, contributing to increase the local chemical heterogeneities and at the same time provide situations that may favour the formation of transverse cracks. These deleterious effects can be reduced if adequate policies are selected, either for the chemical composition of the steel and the process parameters.

4.1.1. MICRO AND MACROSEGREGATIONS

Referring to microsegregations, there are two main differences associated with the continuous casting process of a thin slab with respect to conventional thick slab technology. The first one corresponds to the higher solidification rates that bring about a reduction of the interdendritic space, which is beneficial in terms of microsegregation. The second difference is related to the limitations of working in the peritectic carbon range: the necessity for lower carbon contents will benefit the chemical homogeneity in the solidified structure.

In thin slab casting the solidification takes place in a very short time (approx. 1.5 min) compared to the 15 min (approx.) required for a conventional 200 mm thick slab [153, 154]. These higher solidification rates affect the secondary dendrite arm spacing (λ_{SDAS}). For example, in the region close to the surface of the thin slab, λ_{SDAS} values close to 50 μm have been reported [155, 156]. On the other hand, the shorter solidification times in the centre of the thin slab lead to values in the range of 100-180 μm [157, 158], significantly smaller than those reported in conventional slabs (approx. 250-300 μm). Nevertheless, λ_{SDAS} values higher than 200 μm have been measured in the centre of a thin slab by Ruizhen Wang et al. [159], in the case of low carbon (0.04%) Nb-Ti microalloyed steels.

In this context, it is necessary to consider that λ_{SDAS} also varies with the chemical composition. Won and Thomas [160] proposed the following relationship to quantify λ_{SDAS} as a function of cooling rate (C_R (°C/s)) and carbon content (for the case of % C < 0.15):

$$\lambda_{SDAS}(\mu m) = (169.1 - 720.9 \cdot (\%C)) \cdot C_R^{-0.4935} \qquad \text{Eq. 46}$$

Taking into account the cooling rates obtained in a model predicted by Camporredondo et al. [161] to analyse the thermal evolution during continuous casting (50 mm thin slab with a casting speed of 4.3 mm/min), the values of λ_{SDAS} obtained with the help of Eq. 46 are in the range of those measured by Ruizhen Wang et al. [159] for the case of a 0.04% C. This means that depending on the carbon content, secondary dendrite arm spacing values coarser than those usually assigned to central regions of thin slabs can be obtained. Even so, for a given carbon content λ_{SDAS} will be smaller in thin slab than in conventional slabs, thus, a beneficial effect will be achieved by reducing the scale of segregation [162].

Some continuous casting installations require work outside the hypoperitectic carbon region (0.09 to 0.17% C) in order to avoid longitudinal surface cracks during casting [163, 164] (for example, this is the case of funnel-shaped mould in CSP plants). The peritectic reaction (liquid + δ ferrite transforming into austenite, see Fig. 58) is associated with some shrinkage. This moving away from the mould induces an irregular solidification of the shell, resulting in the formation of longitudinal centred depressions which may produce longitudinal cracks [165] and the strand breakout [2]. This problem means that it is necessary to avoid solidification in the peritectic range. To avoid this region, in the case of conventional slabs, a maximum equivalent carbon content (C_{eq}) of 0.1% is proposed, where C_{eq} is defined as [165]:

$$\mathbf{C_{eq}(\%) = \%C + 0.014\%Mn + 0.023\%Ni - 0.037\%Si - 0.222\%S -}$$
$$\mathbf{- 0.04\%P + 0.003\%Cu - 0.004\%Mo}$$

Eq. 47

In this context, a recent work analysing the influence of alloying elements on the thermal contraction of peritectic steels confirms that, while Mn and V behave relatively neutral, both Cr and Ni have a negative effect on crack sensitivity [166].

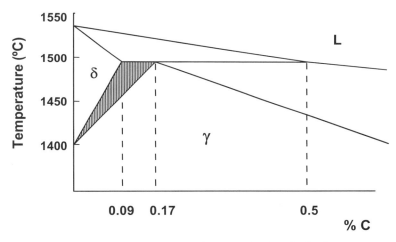

Fig. 58. Equilibrium phase diagram of carbon steel. The dashed zone corresponds to the delta-austenite transformation region.

Nevertheless, the previous condition is not enough to completely assure the absence of longitudinal cracks. The high speeds applied in thin slab continuous casting made the formation of hot spots in the meniscus region critical. These hot spots can appear as a consequence of mould level fluctuations [163]. Their avoidance needs to develop conditions that guarantee a uniform heat removal in the meniscus region. This procedure will guarantee that the delta to gamma transformation takes place in all the zones when the solid shell has sufficient strength and thickness to prevent the formation of longitudinal cracks [163].

The limitation of the carbon content has an indirect beneficial effect on the chemical homogenisation that was initially identified in conventional slabs, but that can be extended to thin slab casts. Subramanian et al. [167] reported that the higher diffusion rates of different substitutional elements in δ ferrite compared to the austenite phase (see diffusion values of different solute elements in Fig. 59 [20, 168, 169]) results in an increased composition homogeneity during the post-solidification process. Consequently, there is a reduction in the intensity of microsegregation of substitutional elements when the selected lower carbon contents lead to a wider post-solidification cooling range in the delta ferrite interval.

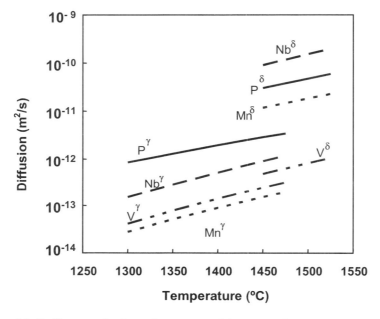

Fig. 59. Diffusion of solute elements in δ ferrite and austenite [20, 168, 169].

To take into account the possible beneficial effects of an increase in the temperature window of the δ ferrite field, through a decrease in the steel carbon content, it is necessary to consider the segregation susceptibility of different elements during the solidification process. The equilibrium partitioning ratio of solute, defined as $k = C_S/C_L$ where C_S and C_L are the solute concentration in the solid and in the liquid, respectively, is shown in Fig. 60 for the case of the main microalloying elements, including also P and Mn. In the figure, an interval is assigned to each element, considering different values published in the bibliography together with results obtained with Thermo-Calc software [160, 162, 170 - 172]. Solute elements with low k values will have the greatest tendency to segregate. According to Fig. 60, P, Nb and Ti are the elements that will most benefit through a solute redistribution during cooling across the delta ferrite region when low C contents are selected. Similarly, as pointed out by Krauss [162], it is necessary to consider also the effect of Mn, which despite its higher value of k compared to the aforementioned elements plays an important role in segregation because of the typical high concentrations of this element in steels.

Another additional benefit of working with carbon contents lower than 0.1% has been pointed out by Han et al. [173]. Based on a microsegregation model, the aforementioned authors predicted, for the case of conventional slabs, that steels containing carbon contents over 0.1% show a significant increase in P segregation. This effect, together with the accumulation of S in the interdendritic liquid, increases the tendency of possible internal cracks. In consequence, the selection of equivalent carbon contents lower than 0.1% should improve the internal thin slab quality.

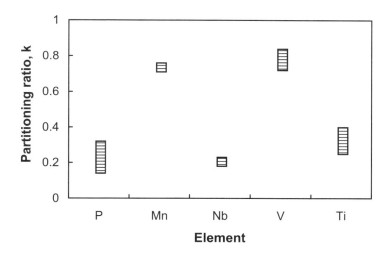

Fig. 60. Equilibrium partitioning ratios [160, 162, 170 - 172].

This increase in composition homogeneity has been related to several improvements in final mechanical properties. Hulka et al. [174, 175] observed that the selection of low carbon contents helps to avoid brittle phase formation and subsequent deterioration of the *HAZ* toughness in welded structures. Similarly, Taira et al. [176] confirmed an improvement in hydrogen induced cracking resistance (through a reduction of intense segregated zones) in line pipes when low carbon steels were selected.

Referring to thin slab macrosegregations, C and S enrichment has been reported only in the proximity of the centreline of the slab in plain C-Mn steels [158]. In this context, the possibility of bulging between rolls during continuous casting needs to be taken into account. This effect occurs when there is a bending effect in the slab between two consecutive rolls, as shown in the scheme of Fig. 61. The origin of this bending can be due to a misorientation in the rolls or to excessive pressure [177]. As a consequence of the bulging there is an enrichment in residual elements, an effect that promotes the formation of a higher density of MnS inclusions in the centre of the thin slab.

This behaviour is not specific to thin slabs, in fact, the central segregation due to bulging was initially identified in thick slabs [178, 179] although the higher casting speeds applied in thin slab casting can enhance it [180]. Its occurrence in thin slabs has been reported [177, 181] and can have detrimental effects in transverse direction properties, mainly in the case of small hot rolling reductions (gauge thicknesses higher that 8-10 mm, for example). It is worth emphasising that the presence of bulging can also have other detrimental effects. For example, Yoon et al. [181] confirmed that bulging in thin slab casting can originate mould level fluctuations, which as mentioned previously, can cause surface defects.

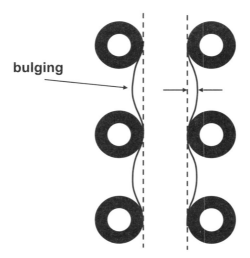

Fig. 61. Scheme of bulging (exaggerated) between two consecutive rolls during continuous casting.

In some plants liquid core reduction is applied. Initially, it was developed in the ISP (In-Line Strip Production) process [1] but later it was also extended to other routes [182]. The objective of this reduction is to start hot rolling with lower thicknesses which allow thinner final gauge coils to be produced and at the same time to increase productivity. Several authors have pointed out that its application could lead to an improvement in the solidification structure reducing the secondary dendrite arm spacing in the centre region of the slab [183] as well as to a minimisation of segregations [184, 185].

4.1.2. TRANSVERSE CRACKS DURING CONTINUOUS CASTING

Transverse crack formation is one of the surface problems which can appear during continuous casting. The short dwell time in the tunnel furnace in TSDR, usually between 15-20 min, compared to conventional reheating furnaces (soaking 2-3 hours at higher temperatures than in the tunnel furnace) reduces the scale jacket thickness (from 0.4 mm in thin slab to 1.5-3.2 mm in conventional walking beam furnaces [3]). This limits the possibility of reducing the surface defects through the removal of the scale layer. Furthermore, as the inspection of the surface before rolling is not possible it becomes more important so as to avoid the formation of transverse cracks during continuous casting.

The origin of these cracks is a combination of tensile strains at very low strain rates together with a significant loss in the steel hot ductility. There are two main situations in which tensile strains act during continuous casting: during bending and straightening operations. In the first case, the lower face of the slab is under tensile strain and the same occurs with the upper face during straightening. When compared to conventional casts, two different aspects require consideration: geometry and cooling rate.

The geometrical aspects related to thin slabs can be considered beneficial in terms of reducing the risks of transverse cracks appearing. The surface strain acting during bending is given by $\varepsilon \approx t/2R$, where t is the slab thickness and R the bending radius [186]. This implies that, for a given radius, the strain value during thin slab bending and straightening will be 3 to 4 times smaller than that present in conventional thick slabs. Similarly, the strain rate is about 5 times higher than in conventional processing [187]. This has been the main reason to assume that thin slabs are less sensitive to hot ductility problems, compared to conventional geometries.

The cooling simulations done by Camporredondo et al. [161] for a 54 mm thin slab (considering a casting speed of 4.3 m/min) showed that, while the surface temperature during secondary cooling oscillated between 900 and 1100°C in the centre of the slab, at the edges temperature falls between 750 to 900°C could occur prior to straightening. These local temperature predictions confirm that, during thin slab continuous casting, the conditions that cause the formation of transversal defects can exist. Although there are few studies published in this field [188], industrial practice reveals that in TSDR care must be also taken. This problem can be more notorious if high nitrogen contents are present, as occurs in steels produced via EAF [187].

Independently of the applied strains, the transverse crack formation is related to a significant decrease in steel hot ductility with temperature, as shown in the scheme of Fig. 62. An abrupt deterioration in ductility occurs when the temperature falls below T_a and the ductility trough remains until the temperature is lower than T_b.

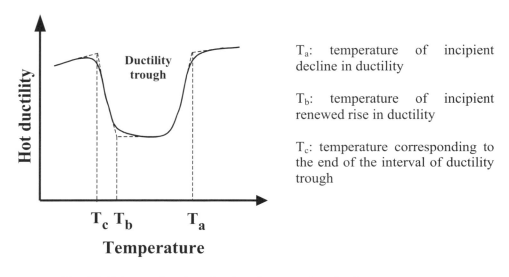

T_a: temperature of incipient decline in ductility

T_b: temperature of incipient renewed rise in ductility

T_c: temperature corresponding to the end of the interval of ductility trough

Fig. 62. Scheme showing the ductility trough at high temperatures.

There are two main microstructural changes associated with the appearance of the ductility trough region: fine particle precipitation and formation of thin films of ferrite surrounding the austenite grain boundaries (see scheme in Fig. 63). In both cases a drop in temperature is required and the cracks are intergranular in nature, as shown in the example in Fig. 64.

In the case of precipitation induced ductility loss, intergranular cracks are enhanced by the formation of precipitate free zones. Strain induced precipitation occurs preferentially at grain boundaries and inside the grains, and it is accompanied by precipitate free zones on both sides of the boundaries (Fig. 63a). These zones are softer than the matrix hardened by a very fine precipitation. When deformation is applied (during bending or straightening operations in continuous casting), strain concentrates at the weak regions promoting microvoid nucleation at the boundary precipitates. Fracture nucleation and propagation takes place via the coalescence of theses microvoids.

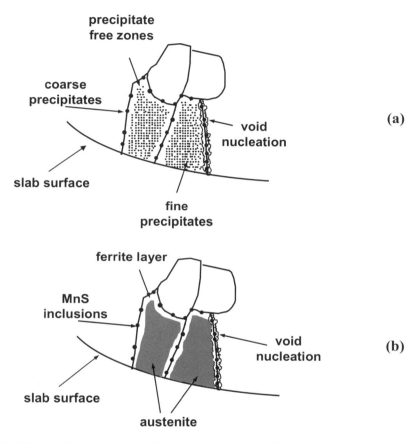

Fig. 63. Scheme showing the nucleation of crack (void coalescence) during bending in the soft regions at: a) precipitate free zones and b) ferrite layers.

In this context, the different solubility and kinetics of elements able to precipitate, under conditions corresponding to continuous casting, need to be evaluated. Fig. 31 showed the solubility curves of the main nitride formers in austenite. In general, Al and Nb are considered the most susceptible elements for the promotion of transverse cracks. Nevertheless, in the case of thin slab casting, B and Ti need also attention. On the other hand, as a consequence of its high solubility in austenite (see Fig. 31), vanadium is unlike to be precipitated in the normal temperature range corresponding to continuous casting.

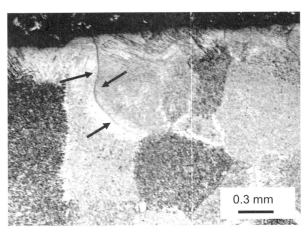

Fig. 64. Transverse crack propagating along austenite grain boundaries in a thin slab. Sample taken at the exit of the continuous casting.

In C-Mn-Al steels, experimental results confirm that increasing either the Al or N levels leads to deterioration in ductility making the trough wider and deeper. Mintz confirms that the value of [Al][N] is the main factor and as a consequence, higher nitrogen levels will increase the risk of transverse cracks being produced [189]. In order to avoid the AlN precipitation that can cause problems during continuous casting, it has been proposed to fix maximum levels of Al ($\leq 0.035\%$) and nitrogen (≤ 90 ppm) [164].

Nb(C,N) precipitation occurs at higher temperatures than AlN and as a consequence, Nb microaddition increases the temperature window of the low ductility region [190]. In this situation, with uneven cooling schedules, the possibility of transverse crack appearance increases. Similarly to Al, in the case of niobium microalloying, low nitrogen contents are recommended [191].

In general, titanium and boron are not considered as likely to promote transverse cracks in conventional thick slabs. Nevertheless, due to the fact that higher solidification rates lead to the formation of finer precipitates, steels with low Ti/N or B/N relationships may become predisposed for transverse cracks to appear in the case of thin slabs [192].

The second mechanism leading to ductility loss is the formation of thin films of ferrite surrounding the austenite grain boundaries (Fig. 63b). During high temperature deformation ferrite is softer than austenite and this makes the strain to concentrate in the ferrite films. When deformation is applied these ferrite films behave in a similar way to the precipitate free zones, leading to the formation of microvoids nucleated at inclusions or precipitates located at grain boundaries. Ductility recovers again when a significant volume fraction of ferrite has been formed on the low temperature side of the trough.

Usually, because of heat losses, slab corners are more inclined to show transverse cracks. In industrial practice, this is avoided by reducing the spray cooling intensity at the corners of the strand. In other cases, transverse cracks can appear sporadically on the upper or lower surfaces. This situation can be caused by excessive local cooling bringing one of the aforementioned microstructural changes: prior to the straightening operations, temperature fluctuations originated by sprays and guide rolls are frequent. These temperature oscillations can enhance precipitation and widen the ductility trough temperature interval.

4.1.3. AS-CAST AUSTENITE GRAIN SIZE

Fig. 65 shows the microstructure of a thin slab quenched at the exit of the continuous casting. Coarse austenite grains are identified in the figure. The metallographic measurements in an industrial thin slab sample in two different zones of the slab are shown in Fig. 66 [193]. One corresponds to the approximately equiaxed grains appearing near the surface and the other belongs to the mixture of coarser equiaxed and columnar grains located in the centre of the thin slab (the grain size has been defined by its equivalent area diameter). From the figure, it can be seen that in the centre of the thin slab there is a significant volume fraction of austenite grains having a size close to 2 mm.

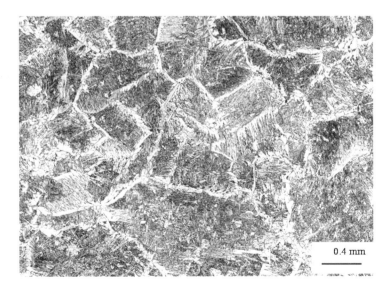

0.4 mm

Fig. 65. Coarse grains in a quenched thin slab at the exit of continuous casting.

This broad grain size distribution is one of the main characteristics in TSDR processes that needs to be considered. It is worth emphasising that the first step during TSDR rolling will require the definition of specific process conditions aimed at reaching a complete refinement of the as-cast microstructure as soon as possible for each steel grade.

Reducing the as-cast austenite grain size could significantly contribute to a simplification of the requirements associated with the initial rolling passes. In order to evaluate possible refinement routes it would be necessary to analyse the different factors intervening in the austenite grain size.

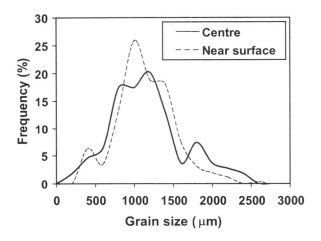

Fig. 66. As-cast austenite grain size distribution in two different zones of a industrial thin slab [193].

The austenite grain microstructure is influenced by the kinetics of solidification, the delta ferrite/austenite phase transformation and finally by the coarsening of the austenite grain size. Pottore et al. considered that the most important factor is the austenite grain coarsening rate [194] and proposed that this step could be controlled by an adequate dispersion of fine precipitates of Ti and Nb. Similarly, Frawley and Priestner [195] observed, in laboratory heats, that austenite grain boundary migration can be effectively suppressed by MnS particles when S levels higher than 0.025% lead to immediate precipitation after δ/γ transformation.

On the other hand, Yoshida et al. [196, 197] evaluated the influence of delaying the δ ferrite/austenite transformation. Taking into account the influence of different elements on decreasing A_{e4} (δ/γ) and increasing A_{e3} (γ/α) temperatures, the aforementioned authors selected phosphorus as the stronger element to stabilise the BCC phase. Yoshida et al. observed that the austenite grain size was reduced by half when the phosphorus content was increased from 0.01 to 0.2%.

In industrial practice, these two different routes to reduce the as-cast austenite grain size are not normally applied. The as-cast grain size controlled by particles remains an open subject which requires significant research and for the majority of steel grades, phosphorus content is limited to a range that its effect in grain size evolution can be considered negligible.

Summarising, the grain sizes shown in Fig. 65 and Fig. 66 represent the typical microstructure for the majority of thin slabs at the entry to rolling mills, such it is confirmed by the results obtained in different plants with several steel grades [198 - 200].

4.2. TUNNEL FURNACE

During the time the thin slab remains in the tunnel furnace, the temperature can affect in the following metallurgical aspects:

— The formation of surface defects.

— Possible fresh precipitation or, in contrast, dissolution of the previously formed precipitates during continuous casting.

— Austenite grain size evolution during rolling, since it determines the rolling start temperature.

The influence of the rolling start temperature on austenite microstructure evolution is considered in the next chapter. The other two aspects are discussed in the following subtasks.

4.2.1. EFFECT OF COPPER AND OTHER TRAMP ELEMENTS IN SURFACE QUALITY

Thin slab surface quality becomes a very important subject when steels are produced following scrap based EAF (electric arc furnace) routes. It should be assumed that tramp elements present in the scrap, combined with the singular conditions current in TSDR, may have a very negative incidence in the slab surface quality prior to rolling. Copper, tin and other residual elements remain as they are not preferentially oxidised and as a consequence build up progressively at the steel/oxide layer. The main problem of this enrichment is hot shortness (the melting point of Cu is 1083°C): Cu segregates along austenite grain boundaries leading to the formation of surface defects during the application of mechanical stresses in subsequent rolling passes.

There are two main features that increase the risks of surface defect appearance in the case of thin slab casting with respect to conventional routes:

– Typical working temperatures in industrial tunnel furnaces are in the range
   of 1040 to 1150°C. This temperature interval can be considered very
   harmful because it coincides with the range at which maximum Cu
   enrichment at the steel/oxide layer and subsequent penetration in the
   austenite grain boundaries occurs [201]. Fig. 67 corresponds to an example
   of Cu penetrating into a grain boundary.

– The detrimental effect of coarse austenite structure on the susceptibility to
   cracking is enhanced since the critical level of Cu enrichment is reached
   earlier than in a finer austenite grain structure (as occurs in cold charging)
   [202].

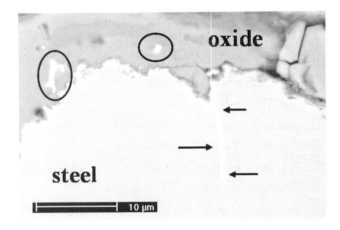

*Fig. 67. Example of Cu segregation in the surface of a thin slab with $Cu_{eq} = 0.15\%$.
There are copper rich globules at the oxide layer (surrounded by a circle) and Cu
segregated at an austenite grain boundary (with arrows).*

The cracking severity of Cu is increased by the presence of other residual
elements, such as Sn, which reduces the Cu solubility in austenite; the reverse
effect is produced by Ni [201, 203]. These opposite effects have been taken
into account defining a Cu equivalent content as [204]:

$$Cu_{eq}(\%) = \%Cu + 10(\%Sn) - \%Ni \qquad \text{Eq. 48}$$

For most applications, the surface quality requirements imply a limitation in the
maximum admissible Cu content. For the case of thin slab casting, Wigman and
Millet proposed a maximum content of 0.15% Cu [205]. Similar maximum
copper contents are used in other CSP plants [206]. It is worth emphasising that
this limit, in agreement with Eq. 48, can be increased by the addition of Ni.

Nevertheless, it is necessary to take into account the fact that Ni addition to Cu containing steels increases the adherence of oxide scale [207]. After descaling this can enhance the presence of remaining spots of oxides prior to hot rolling.

Another possible way to avoid the detrimental effects of Cu is to increase the furnace temperature, this favours the occlusion of Cu rich particles inside the oxide and not at the steel/oxide interface. In Fig. 67 an example of Cu rich globules inside the oxide scale is shown (very oxidising atmospheres also favour the occlusion of Cu in the scale). However, it should be noticed that this could negatively affect the refinement of the final grain size during subsequent rolling, as will be explained in the following chapter.

4.2.2. MICROALLOYING ELEMENTS

Referring to microalloying, there is a series of studies published in the literature that analyse the particle precipitation and/or dissolution phenomena in detail that may take place during the continuous casting and in the following dwell period in the tunnel furnace. This is considered an important factor because, except in the case of Ti, the goal is to have the entire microalloying element content in solid solution before rolling. This is one of the most significant differences when compared to the CCR process, where the use of long reheating times can produce a significant dissolution of microalloying elements.

It is necessary to recognise that the majority of the published data was obtained from laboratory heats, while in a smaller amount of works the analysis has been focused on industrial samples. Although in general, laboratory trials have tried to simulate industrial conditions, the observed discrepancies show that this objective is not always achieved.

The evaluation of possible precipitation/redissolution phenomena of particles before hot rolling will be carried out taking into consideration the following two situations: one concerning single additions of microalloying elements (Ti, V and Nb) and the other multiple microalloying additions.

4.2.2.1. PRECIPITATION OF SINGLE MICROALLOYING ELEMENTS

Concerning single microadditions, vanadium is the element that, owing to its higher solubility (see Fig. 31), is less sensitive to premature precipitation. Only in the case of higher V contents, combined with high N (0.10% V and 0.022% N), little VN precipitation along austenite grain boundaries was observed when low tunnel furnace temperatures (1050°C) were applied [208, 209], some of the precipitates were associated with MnS. In the rest of the cases it can be assumed that this element will remain completely in solution [210].

With Nb microalloying more care must be taken with this element in order to avoid premature precipitation. Park et al. [211] considered the influence of C content in different Nb microalloyed steels (laboratory heats). They observed that for C > 0.07% some Nb segregation occurred during the final stages of solidification leading to the formation of coarse eutectic Nb(C,N) precipitates, resulting in a loss of about 25% of the total Nb for further precipitation. This type of particle has not usually been reported in industrial sample analysis which could confirm that they may be a consequence of the peculiar segregation conditions in the laboratory trials. Nevertheless, these results also confirm the above mentioned relevance of having low C contents so as to favour homogenisation during solidification and subsequent cooling.

With the help of a precipitation model and experimental analysis, Jacobs et al. [212, 213] identified the importance of limiting the nitrogen content in order to avoid Nb(C,N) precipitation during the equalising time in the furnace. El-Bitar et al. [191] defined, based on this work, what the maximum nitrogen content should be if Nb(C,N) precipitation during casting and in the tunnel furnace needs to be suppressed at a given tunnel furnace temperature (Fig. 68).

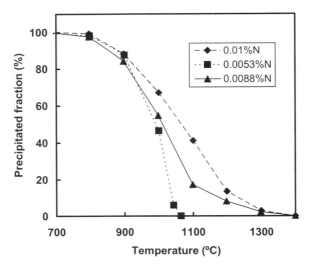

Fig. 68. Influence of nitrogen on Nb(C,N) precipitation in the case of 0.06% C and 0.03% Nb steel (adapted from ref. 191).

In relation to titanium, Nagata et al. [214] analysed commercial thin slab steels with Ti contents between 0.009% and 0.048% and Ti/N relationships between 0.6 and 6. They observed differences in the precipitation behaviour depending on the titanium and nitrogen contents. On the other hand they also observed that precipitation was finer and more homogeneous than in thick slabs. In the as-cast conditions they measured mean precipitate sizes smaller than 20-30 nm.

After the tunnel furnace step, at 1100°C for 30 min, larger particles were observed whose mean values between 40 and 50 nm. This increase in size was more important in the case of hyperstoichiometric Ti/N relationships. In some of the steels, the authors suggested that TiN precipitation might be incomplete after tunnel furnace soaking, bringing about the possibility of strain induced precipitation or, depending on the chemical composition, particle coarsening during rolling. The presence of fine particles formed during rolling was previously observed by Gibbs et al. [215] in a 0.02% Ti (0.005% N) laboratory heat TSDR simulation.

Moreover, the dimensions of the TiN particles found in thin slabs are in the range of sizes considered as adequate to pin the austenite grain boundaries and thereby inhibit grain growth after recrystallisation during rolling (recrystallisation controlled rolling). Similarly, in a recent work it was observed that the smaller size of the precipitates formed prior to hot rolling in near-net-shape casts compared to conventional thicker ones, could also exert some delay in recrystallisation kinetics [77, 216].

It should also be pointed out that because of the high post-solidification cooling rates associated with thin slab casting, very fine precipitation of different residual elements may occur (as for example when Ti is present as residual element coming from scrap). These small particles may also exert a pinning effect on austenite grain boundaries limiting grain growth. For example, an industrial TSDR plain C-Mn steel with a mean grain size as small as 65 μm was measured [217] after reheating the sheet for 30 min at 1300°C. In Fig. 69 the grain size evolution of a TSDR sheet is compared with that shown by a conventionally processed steel. Grain coarsening is significantly higher in the latter. This difference was attributed to the pinning effect produced by very fine TiN precipitates formed during solidification and post-solidification by the residual Ti present in the scrap. Furthermore, these fine particles can also have a positive effect during subsequent welding operations, since the absence of grain growth contributes to an improvement in the mechanical properties of the heat affected zone.

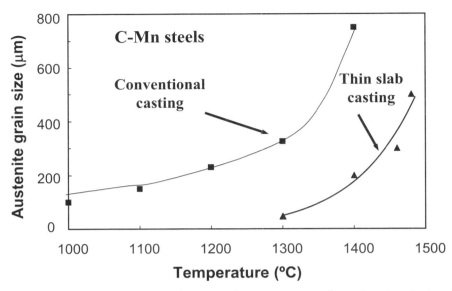

Fig. 69. Austenite grain size evolution with temperature after reheating during 30 min in two plain C-Mn steels. The steel produced via TSDR shows a significant delay in grain growth [217].

Finally, in this context the fine precipitation of MnS and CuS should be considered. In direct charging the degree of supersaturation with respect to these sulphides may be significantly greater at the same rolling temperature as in cold charging. As a consequence, precipitation can occur during hot rolling (strain induced precipitation) and/or during transformation [195, 218 - 221]. Their very small size can affect austenite softening mechanisms during rolling and also increase final room temperature strength.

4.2.2.2. PRECIPITATION OF MULTIPLE MICROALLOYING ELEMENTS

The situation when multiple microalloying elements are present becomes considerably more complex. In this field, the different works done by Li et al. must be emphasised [208, 209, 222, 223]. The main results of these works can be summarised as follows:

—   V-Nb steels (0.11% V, 0.030% Nb, 0.011% N): precipitation was not observed after casting. In the tunnel furnace, at a temperature of 1100°C complex cuboid particles of size 50-110 nm, containing both V and Nb, were found [222]. Thermodynamic calculations confirmed that the solution temperature of these precipitates was around 1158°C.

– Ti-Nb, Ti-V and Ti-V-Nb steels: the introduction of titanium significantly changes the precipitation behaviour. Titanium in V, Nb and V-Nb bearing steels increases the dissolution temperature of the carbonitrides, and as a result, complex precipitates with different geometries appear both during continuous casting and in the tunnel furnace [208, 209, 224, 225]. Some examples of Ti precipitated in a Ti-V-N laboratory heat [226] and in a Ti-Nb-V industrial thin slab [159] are shown in Fig. 70.

In the case of (Ti, Nb)(C,N) particles, while Li et al. [209] observed some additional precipitation during the equalisation period, Garcia et al. [224] confirmed that some dissolution took place at a temperature of 1150°C in the tunnel furnace.

On the other hand, in the case of Ti-Nb steels, a reduction in Cu content was proposed, as was observed that coprecipitation of (Ti,Nb)(C,N) could occur on previously formed CuS particles [212].

The loss of microalloying elements caused by premature precipitation before rolling, mainly in the case of Ti additions, leads to some reduction in yield strength (compared with heats without Ti) although at the same time an improvement in toughness is reported [222, 223]. Those situations where early precipitation takes place (when Ti is combined with V or Nb), do not seem to originate microalloying element losses larger than those observed in conventional CCR processes [114].

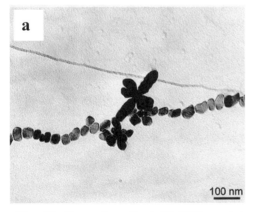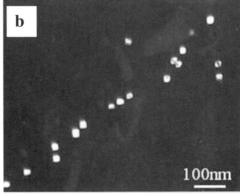

Fig. 70. Examples of: a) (Ti,V)N particles precipitated in a Ti-V-N laboratory heat [226] and b) TiN particles in a Ti-Nb-V industrial thin slab [159].

In summary, by suitable controlling the chemical composition (C and N, mainly) and the equalisation temperature in the tunnel furnace, in most situations it will be possible to use of all the potential of any microalloying elements in subsequent steps during the process. Under these conditions (mainly for the case of Nb and Ti microalloyed steels), higher supersaturation levels than those obtained in CCR can be achieved. This will provide a larger driving force for precipitation to occur at higher temperatures and also for higher solute drag effects in retarding recrystallisation during rolling [93].

5. TSDR: ROLLING OF PLAIN CARBON AND MICROALLOYED STEELS

The metallurgical peculiarities described in the previous chapter define a situation that is differentiated from the point of view of thermomechanical processing. It is well known that the main objective of thermomechanical processing is the conditioning of the austenite before transformation. In the case of conventional controlled rolling this is based on austenite refinement followed by strain accumulation. Similarly, in TSDR two different steps should be considered when defining rolling sequences:

-   Complete elimination of the initial as-cast microstructures in the first rolling stands.

-   Once this objective is achieved, optimum austenite conditioning prior to transformation.

It must be taken into account that the total reduction that can be applied in the 5-7 rolling passes in TSDR is significantly smaller than that applied in the conventional route. This means that the thermomechanical parameters should be optimised to achieve the recrystallisation of the as-cast microstructure as soon as possible. The remaining rolling passes should provide the adequate austenite conditioning prior to transformation.

5.1. AS-CAST AUSTENITE REFINEMENT MECHANISMS

The refinement of the as-cast microstructure involves the activation of softening mechanisms. Taking into account the fact that the static recrystallisation kinetics depend on the initial austenite grain size, the refinement of the as-cast grains will be more complex than in the case of cooled (transformed) and reheated finer austenitic microstructures corresponding to conventional cold charge rolling (200-300 μm).

Fig. 71 shows the recrystallisation time, calculated for different initial austenite grain sizes normalised by the time calculated for a grain size of 200 μm (t_{200}) plotted against the austenite grain size for different applied strains. For time calculations the equation previously shown in Table 9 for Nb and Nb-Ti steels [74], applicable to a wide range of initial austenite grain sizes, was employed. From the figure it can be observed that for deformations between 0.4 and 0.7 (which are typical values in the first stand of a CSP plant), a change in the grain size from 200 μm to 1 mm increases the recrystallisation time between 3 and 4 times, respectively. For a grain size as coarse as 2 mm the time required would be double the previous one. It is worth emphasising that, as is shown in Fig. 66, in a real situation a significant fraction of grains belongs to the latter group.

Some authors have proposed that the activation of dynamic recrystallisation could induce a faster microstructural refinement [227, 228]. This hypothesis is based on the assumption that the large reductions applied during the initial rolling passes in the TSDR process could exceed the critical strain for the onset of dynamic recrystallisation (ε_c) and subsequently activate metadynamic softening. This could have a positive effect and account for the significant microstructural refinement related to these processes, due to the fact that both dynamic and metadynamic grain sizes are independent of the initial austenite grain size. Nevertheless, there are two aspects that should be considered concerning the initial coarse grain size distribution, as they make the above softening mechanism difficult to operate. The first is related to the beginning of dynamic recrystallisation and the second to the conditions required for metadynamic recrystallisation (MDRX) to operate as the main post-dynamic softening mechanism.

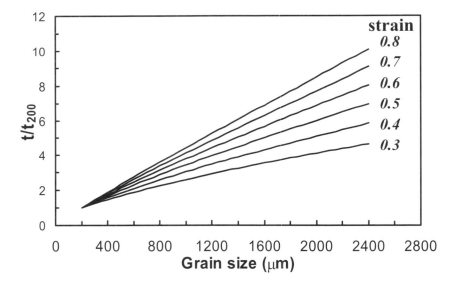

Fig. 71. Influence of initial austenite grain size and applied strain on recrystallisation time, the latter normalised by the t_{200} (recrystallisation time corresponding to a grain size of 200 μm).

Regarding the onset of dynamic recrystallisation, Fig. 72 shows the dependence of the critical strain ε_c on the austenite grain size and temperature [229], in the case of a 0.035% Nb microalloyed steel deformed at a strain rate of 5 s^{-1}, deformation conditions that are similar to industrial conditions in the first rolling stand (the equation included in Table 8 for Nb and Ti-Nb steels was considered [65]). For typical industrial rolling conditions, the Zener-Hollomon parameter ranges from 10^{13} to 10^{14} for the first pass, from 10^{14} to 10^{15} at the second rolling stand and finally, from 10^{15} to 10^{16} for the third stand. These ranges are indicated by the shaded areas in Fig. 72 as a reference.

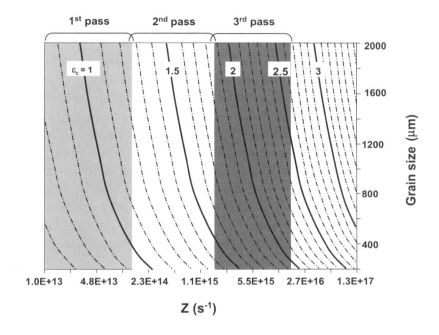

Fig. 72. Critical strain ε_c for the onset of dynamic recrystallisation as function of temperature and prior grain size (0.035% Nb, strain rate 5 s^{-1}) [229].

For the first rolling pass and Z values around $5 \cdot 10^{13}$ (corresponding to a strain rate of 5 s^{-1} and a rolling temperature of 1100°C), the application of a strain $\varepsilon = 0.8$ would activate dynamic recrystallisation for grains smaller than 200 μm, though an increase of strain to $\varepsilon = 1$ would be necessary for the activation of dynamic recrystallisation in the grain size interval up to 1 mm. This clearly denotes the convenience of applying larger reductions in the first pass. Even so there will be an important fraction of grains, larger than 1 mm, which would only start recrystallising after the pass is finished, i.e. under static recrystallisation conditions. Hence, it is really difficult to overcome the critical strain for dynamic recrystallisation in the whole size range of the coarse as-cast austenite.

If the typical conditions for the second rolling pass are considered (lower temperatures and higher strain rates) the critical strain levels substantially increase, as indicated in Fig. 72. The same occurs for the third pass. This means that if conventional strains per pass are considered only the grains that remain unrecrystallised after the first interpass interval, thus retaining a certain level of deformation, would be under dynamic recrystallisation conditions after subsequent passes. This point may be important for schedule design. Although very high levels are required for *DRX* to activate in the case of the coarsest grains, the retained strain from one pass to the next could help reaching the required ε_c values at the second or third stand.

Secondly, it is usually assumed that when $\varepsilon > \varepsilon_c$, post-dynamic softening is governed by metadynamic recrystallisation. Nevertheless, as shown in the scheme in Fig. 73, under those conditions where there are big differences between the initial austenite grain size, D_o, and the dynamically recrystallised grain size, D_{dyn}, the characteristics of the metadynamic recrystallisation are only achieved after some minimum strain, ε_T, is reached. This strain is denoted as the transition strain and is related to the minimum amount of dynamically recrystallised fraction (see Fig. 19) guaranteeing that MDRX can account for a complete softening, assuming that this process only involves growth of previous dynamic grains. It should be pointed out that, in contrast to what was expected for coarse austenite grain sizes, intragranular nucleation of dynamically recrystallised grains at twins or other intragranular defects did not significantly contribute to microstructural refinement [64].

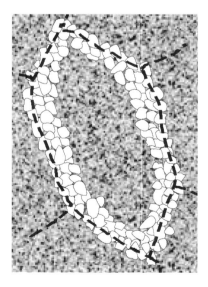

Fig. 73. Example of big differences between a coarse initial austenite grain size and small DRX grains.

The strain ε_T may be significantly larger than ε_c. A relationship of $\varepsilon_T = 2.2\ \varepsilon_c$ has been proposed [73] for the case of Nb microalloyed steels with coarse initial austenite microstructure. Consequently, there should be a transition between the strain range where static recrystallisation operates as the post-dynamic softening mechanism ($\varepsilon < \varepsilon_c$) and the strain required for metadynamic recrystallisation ($\varepsilon > \varepsilon_T$).

In industrial processing conditions, it will be really difficult to exceed ε_T and most situations will correspond to either fully classical static recrystallisation ($\varepsilon < \varepsilon_c$) or to mixed conditions (static and metadynamic, $\varepsilon_c < \varepsilon < \varepsilon_T$). The different circumstances that can be given are shown schematically in Fig. 74 [65].

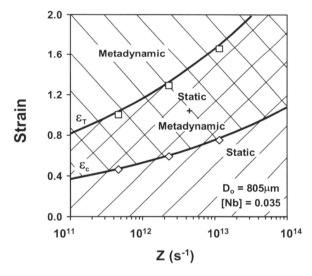

Fig. 74. Static, metadynamic and mixed combination ranges as function of ε and Z
for $D_o = 805\ \mu m$ and $Nb = 0.035\%$ [65].

Another aspect to be taken into account is the kinetics when both static and metadynamic softening mechanisms operate during the interpass interval. Fig. 75 shows the dependence of the 95% recrystallisation time ($t_{0.95}$) with strain, calculated for the case of a 0.05% Nb steel, deformed at 1100°C, a strain rate of $\dot{\varepsilon} = 5\ s^{-1}$ and $D_o = 1000\ \mu m$, taking into account the equations of Table 9 and Table 10 [74, 73]. The figure clearly shows that very high strains have to be applied to assure fast recrystallisation rates controlled by metadynamic recrystallisation.

As mentioned above, the application of a strain higher than the critical value (ε_c) does not imply that metadynamic recrystallisation is thoroughly extended and therefore a transition range of strains may be observed, where static and metadynamic recrystallisations combine. Finally, the longer times required for a complete softening at the conditions where pure static recrystallisation occurs are also obvious. Therefore, even for deformation conditions with a strain larger than the critical one, recrystallisation kinetics would not be as fast as predicted by metadynamic events. For a complete recrystallisation to be reached, the interpass time should be longer than the $t_{0.95}$ values plotted in Fig. 75.

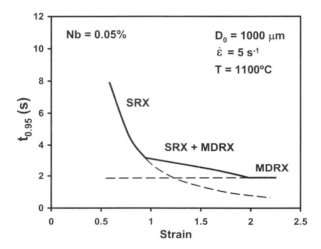

Fig. 75. Dependence of $t_{0.95}$ (time for 95% recrystallisation) on strain on different softening ranges [229].

In most of the industrial thin slab direct rolling plants, all the deformation is applied in a total of 6-7 passes. The previous results suggest that a proper definition of dummy passes (by the concentration of strain at some of the initial rolling stands) may become a very useful tool to achieve complete recrystallisation during rolling. Once the critical strain is overcome, a certain time is needed for complete recrystallisation and the prolongation of the interpass time by dummy passes can be very helpful to accelerate microstructural refinement.

5.1.1. C-Mn AND C-Mn-V STEELS

The reductions usually applied in the first stands are large enough to achieve a complete recrystallisation of the as-cast microstructure at the end of the first or second interstand in the case of plain carbon steels. Only when low initial temperatures and small reductions are applied in the first stand do the coarsest grains remain unrecrystallised, as shown in Fig. 76. In this figure the evolution of the time required for a complete static recrystallisation in the case of a plain C-Mn steel is drawn as a function of the initial austenite grain size.

If an interpass time of ~ 5 s is considered, under those conditions corresponding to low initial rolling temperatures (below 1060°C) and small reductions (less than $\varepsilon = 0.5$) some fraction of coarse grains will remain unrecrystallised when entering the second pass. In these cases, the accumulation of strain during the first interstand will help to complete recrystallisation during the second interstand.

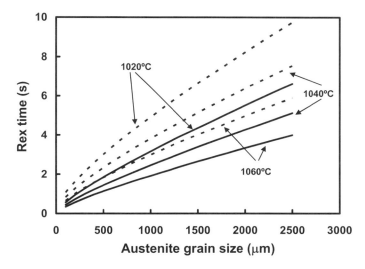

Fig. 76. Time required for complete static recrystallisation in isothermal conditions as a function of the initial austenite grain size for a strain rate of 5 s⁻¹. Full lines correspond to ε =0.5 and dotted lines to ε = 0.4 (plain C-Mn steel).

When microalloying is introduced both solute drag effect and the possible interaction with early strain induced precipitation phenomena need to be evaluated. In the case of vanadium microalloyed steels the effect of this element as a solute drag recrystallisation delaying factor is very small (see Fig. 24) and its behaviour during initial rolling stands could be considered as that corresponding to a plain C-Mn steel [83]. Things significantly change in the case of Nb microalloying. This will be analysed in next section.

5.1.2. Nb MICROALLOYED STEELS

The combination of the relatively large solute drag effect exerted by Nb, with recrystallisation being retarded owing to the coarse as-cast microstructure, introduces important differences compared to conventional controlled rolling processes. On the other hand, in relation to static recrystallisation/precipitation interaction, the occurrence of early Nb(C,N) precipitation could have a very negative effect on the refinement of the as-cast microstructure. If precipitation occurs before this refinement has been completed, the process may be completely halted.

An evaluation of the Nb solute drag effect on the static recrystallisation kinetics is illustrated in Fig. 77, based on the empirical expression [74] listed in Table 9. The delay is relatively small at high temperatures, compared to the behaviour of a plain C-Mn steel. However, when the temperature decreases the difference significantly increases, manily for the highest Nb content.

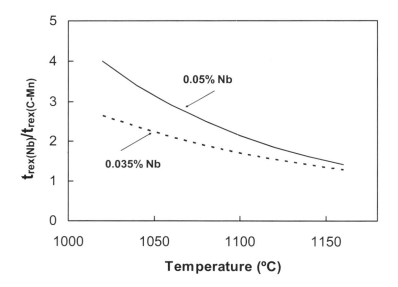

Fig. 77. Delay in static recrystallisation time ($t_{0.5srx}$) due to Nb solute drag effect normalised to the time corresponding to a plain C-Mn steel.

When recrystallisation of the as-cast microstructure is not complete, the final microstructure is characterised by the presence of local heterogeneities, inherited from the unrecrystallised coarsest as-cast grains. Fig. 78 shows two cases where incomplete recrystallisation of the as-cast microstructure has induced detectable heterogeneities in the final microstructure after transformation. These correspond to relatively large areas that transform to larger ferrite grains or, in the case of faster cooling rates, low transformation temperature phases, such us bainite or acicular ferrite.

As it was explained in section 2.4, these microstructural features are often related to a decline in the final mechanical properties, mainly in terms of toughness. An example of a very coarse cleavage facet associated to one of these microstructural heterogeneities is shown in the fractograph of Fig. 79.

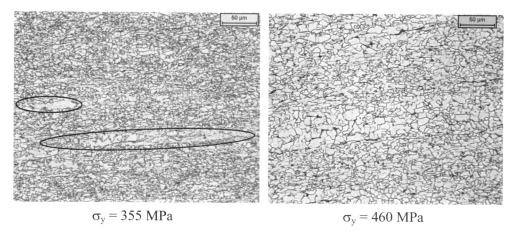

$\sigma_y = 355$ MPa $\qquad\qquad\qquad$ $\sigma_y = 460$ MPa

Fig. 78. Room temperature microstructural heterogeneities in hot rolled strips.

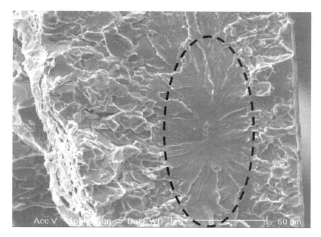

Fig. 79. Coarse cleavage facet (approx. 100 μm length)

A problem that arises when these local heterogeneities are present in the microstructure is that they cannot be correctly identified if the microstructure is analysed in terms of the mean grain size. For example, Fig. 80 shows the austenite grain size measurements compared to predictions from a model based on mean grain size values applied to a laboratory simulation in a Nb microalloyed steel, whose initial mean grain size was ~ 800 μm and which was deformed by six passes starting at 1100°C and finishing at 914°C [230]. The model predicts an important refinement in the final austenite grain size, but does not provide information about the possible presence of residual unrecrystallised coarse initial grains, which are frequently observed experimentally after the TSDR process.

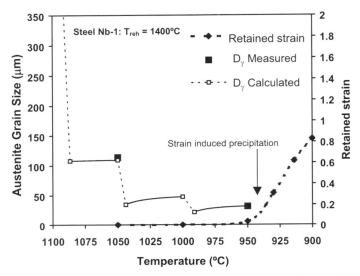

Fig. 80. Evolution of austenite grain size and retained strain during deformation of a 0.035% Nb microalloyed steel ($D_o = 800$ μm, deformed by six passes from 1100 to 914°C, $\varepsilon_{1,2,3} = 0.55$, $\varepsilon_{4,5} = 0.31$ and $\varepsilon_6 = 0.21$ at a strain rate of 5 s^{-1}) [230].

5.1.3. MODELLING OF EVOLUTION OF GRAIN SIZE DISTRIBUTIONS

In order to predict possible microstructural heterogeneities which could impair the final mechanical properties, a model which incorporates the wide range of austenite grains sizes present in the initial as-cast microstructure was developed in [231]. The objective of the model is for it to become a tool for designing thermomechanical processes applied to thin slab direct rolling.

The model considers the influence of coarse austenite grain sizes and high supersaturation levels of microalloying elements on the recrystallisation and precipitation kinetics. The model incorporates the initial austenite grain size distribution as input, instead of the mean value, and considers the following mechanisms interacting simultaneously: static and metadynamic recrystallisation, grain growth during the interpass time and Nb(C,N) strain induced precipitation. With the introduction of these peculiarities the model predicts reasonably well, for a given chemical composition, the processing conditions where an important fraction of coarse unrecrystallised austenite grains remain in the final microstructure prior to transformation.

The initial two dimensional (2-D) austenite grain size distribution is converted into a three dimensional one (3-D), with the help of the method developed by Matsuura and Itoh [232] for any grain size distribution type. The resulting distribution is divided into n intervals (usually between 10 and 15), each one defined by its mean grain size and the corresponding volume fraction (calculated assuming a spherical geometry). The interval (i (=1÷n)) is considered as a homogeneous material and its microstructural evolution during hot rolling is calculated by the semi-empirical equations described in Chapter 3. A scheme showing these steps is illustrated in Fig. 81.

At the end of the first interstand time the recrystallised and unrecrystallised volume fractions are considered independently. The following data are obtained for each i interval: the recrystallised fraction, $[X]_i$, and grain size, $[D_r]_i$, the unrecrystallised grain size, $[D_u]_i$, and the retained strain, $[\varepsilon_r]_i$, in the unrecrystallised fraction.

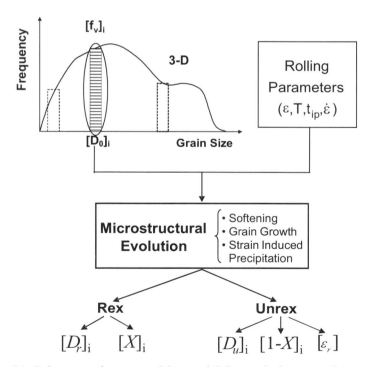

Fig. 81. Schematic diagram of the model for each discretised interval.

On the other hand, for each *i* interval, a grain size distribution of recrystallised grains was considered instead of the mean value. To build these individual distributions, the following two aspects were taken into account:

–   The grain size distributions of recrystallised microstructures can be properly described by log-normal distributions [233].

–   Experience shows that for the case of recrystallised microstructures the maximum/mean grain size ratio stays between 3 and 4 [234].

Consequently, a log-normal distribution of recrystallised grains was derived from each initial grain size interval. These log-normal distributions were built using a mean value equal to the one calculated, $[D_r]_i$, and a maximum grain size 3 times larger than the mean one. Once this modification is introduced in the model, the overall recrystallised microstructure is obtained again by the composition of all the resultant log-normal grain size distributions of the *n* intervals weighted by its corresponding initial volume fraction.

In the case of a complete recrystallisation, the microstructure will be defined by a unique distribution of recrystallised grains; for partially recrystallised conditions, two distributions of grains, one for recrystallised and another for the unrecrystallised ones, with their corresponding fractions will be required.

The results obtained at the end of one interpass period are considered as the input for the next rolling pass. Each data is analysed separately, proceeding as in the first pass and taking into account the fact that the unrecrystallised material retains all the deformation applied in the previous pass. This procedure is repeated throughout the rolling sequence, as shown in Fig. 82.

The application of the model to different processing conditions (Nb content, initial rolling temperature, rolling schedule,...) allows an independent evaluation of the influence of each parameter on the recrystallisation kinetics. This becomes a very useful tool for finding out how to proceed in order to reach a complete refinement and subsequent conditioning of the austenite prior to phase transformation.

Fig. 83 shows an example of the application of the model. It shows the austenite grain size distributions obtained at the end of the third interstand in a 0.035% Nb microalloyed steel, rolled to a final gauge thickness of 8 mm (initial slab thickness of 55 mm) as a function of the initial rolling temperature [193]. The as-cast grain size distribution of Fig. 66 has been considered as input. The strain rates and interpass times selected correspond to typical values of CSP industrial plants.

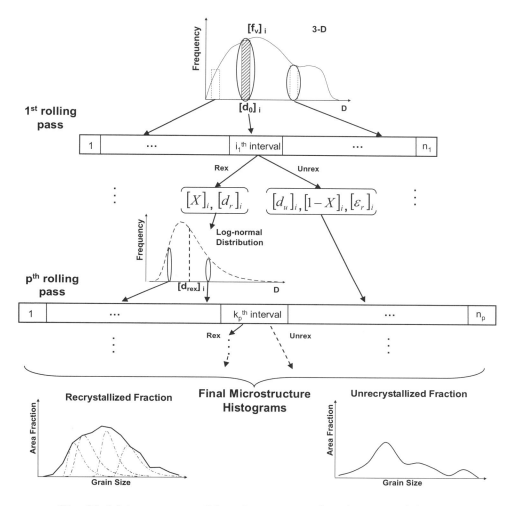

Fig. 82. Main structure of the microstructural evolution model [235].

If low temperatures are selected, the model predicts that an important fraction of unrecrystallised coarse grains remains in the microstructure after the third pass. The model also gives as additional information that this behaviour is caused by early Nb(C,N) precipitation which completely stops the recrystallisation of the aforementioned grains. Therefore, the problem of these residual unrecrystallised grains gets worse when premature strain induced precipitation takes place.

In Fig. 84 the volume fraction of the as-cast microstructure that remains unrecrystallised after each of the first three interstands is drawn as a function of the initial rolling temperature, considering rolling schedules corresponding to an initial thickness of 55 mm and final gauges of 2.5 and 5 mm, respectively

[236]. The rolling parameters selected are listed in Table 15. Under these conditions, the model predicts that as the initial rolling temperature decreases the unrecrystallised as-cast austenite fraction that remains at the entry of the second stand increases. In the same figure, the volume fraction of unrecrystallised austenite due to the effect of strain induced precipitation is also represented.

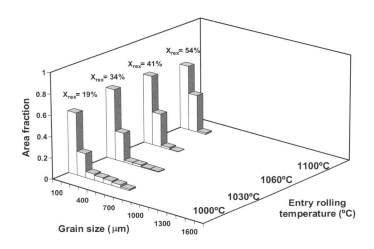

Fig. 83. Prediction of austenite grain size distribution at the end of the third interstand as a function of the initial rolling temperature (0.035% Nb, $\varepsilon_1 = 0.8$, $\varepsilon_2 = 0$, $\varepsilon_3 = 0.4$ for a final gauge thickness of 8 mm) [193].

Table 15. Rolling parameters considered in the computer simulations.

Pass number	Strain		Strain rate (s^{-1})	Interpass time (s)	ΔT (°C)
	e = 2.5 mm	e = 5 mm			
1	0.85	0.60	5	6	35
2	0.80	0.60	10	4	30
3	0.70	0.55	15	3	30
4	0.50	0.40	25	2	30
5	0.40	0.35	50	1	30
6	0.30	0.25	75		

It must be pointed out that those areas that have completely recrystallised after the first pass but not after the second one because of precipitation, are also contributing to the latter value. In all cases, at the end of the first interstand, there is no strain induced precipitation and, consequently, the delay in the static recrystallisation is only attributed to a combined effect of coarse initial austenite grain sizes and Nb solute drag. At the entry of the third stand recrystallisation is completed only when high initial rolling temperatures and large reductions are applied (e = 2.5 mm). In the rest of the cases a certain amount of unrecrystallised as-cast austenite fraction remains.

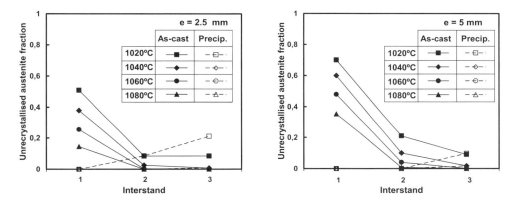

Fig. 84. Evolution of the unrecrystallised as-cast austenite volume fraction after each first three rolling interstands. The volume fraction that remains unrecrystallised by the effect of strain induced precipitation is also included (0.035% Nb steel) [236].

For the lowest initial rolling temperature of 1020°C, the model predicts that precipitation is taking place at the first rolling stages. These precipitates interact with recrystallisation, stopping it and the result of this is that some of the as-cast deformed coarse grains stay unchanged during subsequent rolling passes. In contrast, when higher initial rolling temperatures are selected the precipitation is delayed, thus, it does not affect the initial microstructural refinement.

Some small differences can be appreciated between the microstructural evolutions during rolling for the schedules corresponding to e = 2.5 and 5 mm gauges. In the former case, as a result of the larger strains applied in the first two rolling passes a nearly complete recrystallised microstructure can be reached at the entry of the third rolling pass (for $T_i \geq 1040$°C), whereas, for the case of e = 5 mm, at 1040 and 1060°C the third pass is required for a complete refinement. As expected the application of smaller reductions, for thicker final gauges, retards the completion of recrystallisation of the as-cast microstructure.

This means that for a complete refinement of the as-cast microstructure a minimum initial rolling temperature is required. This temperature will depend on the chemical composition (Nb in solution) and rolling schedule, but its value must be selected in order to assure that strain induced precipitation will not start before the refinement of the as-cast microstructure has been completed.

5.2. AUSTENITE CONDITIONING

The previously described as-cast microstructural refinement must be coupled with a proper conditioning of the austenite before transformation by using adequate rolling parameters in the subsequent stands. In TSDR similar thermomechanical processing approaches to the ones used in classical cold charging can be considered; with recrystallisation controlled rolling (mainly by Ti microalloying) and conventional controlled rolling (retained strain in the austenite) being the two that are most widely applied.

In CCR the main mechanism to stop recrystallisation and develop a pancaked austenite microstructure is the interaction of recrystallisation with strain induced precipitates, as indicated in Section 3.2. In this case, the most suitable microalloying element to achieve this objective is Nb. However, when the effect of Nb microalloying in the TSDR process is analysed some differences are revealed.

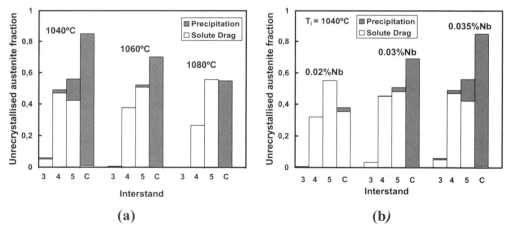

Fig. 85. Unrecrystallised austenite volume fraction due to solute drag and strain induced precipitation (C denotes cooling interval between the final rolling pass and 800°C) for a deformation schedule with e = 2.5 mm: a) effect of initial rolling temperature (for a 0.035% Nb steel) and b) influence of Nb content (T_i = 1040°C) [236].

For example, Fig. 85a illustrates the predictions of the aforementioned microstructural model for the evolution of the unrecrystallised fraction in the last four interstands (selected rolling schedule to obtain a final gauge of 2.5 mm shown in Table 15) [236], considering three different initial rolling temperatures in the case of the same 0.035% Nb steel. With the help of the model, it is possible to separate the contribution of solute drag and strain induced precipitation mechanisms. In the figure, the austenite state at 800°C has been included (denoted by C), assuming a post-rolling cooling rate of 19°C/s.

The model predicts that the solute drag effect is the main mechanism acting to avoid recrystallisation during the fourth interpass interval, with a major effect for the case of the lower initial rolling temperature (higher unrecrystallised fraction). During the next interstand it can be seen that precipitation only becomes active for the 1040°C entry rolling temperature. In the other two cases, strain induced precipitation is irrelevant between rolling passes and only appears during the final cooling. After the final pass, this mechanism is the most important for avoiding recrystallisation.

The influence of precipitation on each interstand can significantly change with the niobium content in the steel. In Fig. 85b this effect has been evaluated taking into consideration three different Nb contents. The precipitation gains relevance before the final rolling pass for a 0.035% Nb, but the increase in Nb shows its greatest influence during the subsequent cooling after rolling. The results corresponding to 0.02% Nb also indicate that some minimum precipitation is required in order to maintain (or increase) the accumulated strain achieved by solute drag. If precipitation does not start during cooling, the solute drag effect cannot completely stop recrystallisation during the time interval between the last stand and 800°C.

The previous results indicate that the solute drag mechanism also plays an important role in delaying recrystallisation during the final rolling stages. This effect depends on the initial rolling temperature. If this temperature is very high, recrystallisation delay is not so effective and grain growth during the interpass time becomes more significant. This means that to optimise the beneficial effect of solute drag in pancaking the austenite during final rolling passes, the initial rolling temperature must be properly selected [237]. On the other hand, strain induced precipitation appears at latter stages of rolling, and mainly during cooling after the last pass for the range of Nb contents considered.

These predictions can be confirmed with the help of the roll forces measured in industrial conditions. In Fig. 86 the mean flow stress values determined from industrial rolling forces have been compared to the predictions of the Shida equation [128], for a plain C-Mn steel and a Nb microalloyed steel, both rolled following the same schedule in order to obtain a final gauge of 3 mm (in the figure the values corresponding to the last four passes have been considered). In the case of the Nb microalloyed steel, there is a clear increase in the mean flow stress in the final two rolling passes, meaning that the microstructure is partially recrystallised, as predicted by the model.

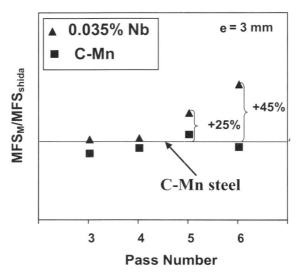

Fig. 86. Example of increase in the mean flow stress, determined from industrial rolling forces, in the last passes for the case of a 0.035% Nb microalloyed steel.

From the above it can be concluded that solute drag exerts a relevant role to obtain strain accumulation in the TSDR process, in contrast to what is usually accepted in traditional routes, where strain is mainly retained as a consequence of precipitation. In this context, if less Nb is precipitated during hot rolling, a higher fraction of this element will be available to increase the strength via different hardening mechanisms associated with the transformation [238].

Finally, if solute drag becomes an important mechanism to obtain pancaked austenite, other microalloying elements can also be considered for that purpose. For example, this idea has been industrially developed for the case of V-Mo combinations [239]. From the analysis of the roll separating forces Chiang pointed out that strain is accumulating in the final rolling stands in V-Mo microalloyed steels for final gauges close to 3.2 mm, as is shown in Fig. 87.

In this case, because of the lower effect of these elements compared to Nb, higher microalloying concentrations are required and in general, this mechanism should be considered as a complement to other main strengthening routes assigned to vanadium (usually, precipitation hardening). Similarly, Nb-Mo combinations could be considered, mainly in those situations where Nb content needs to be limited to avoid premature strain induced precipitation.

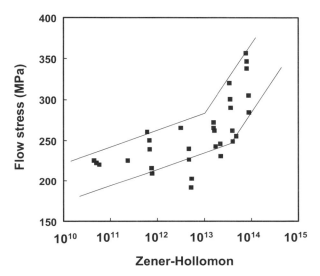

Fig. 87. Mean flow stress as a function of the Zener-Hollomon parameter in the case of several industrial trials of V-Mo microalloyed steels (e = 3.2 mm) (adapted from ref. [239]).

5.3. PROCESSING MAPS

In industrial conditions, the adequate selection of processing parameters, for each specific chemical composition and final gauge thickness, can be complex because of the important number of variables intervening simultaneously. For example whereas to achieve a thin final gauge (thinner than 3 mm, for example) the use of a high initial rolling temperature can result in some growth of recrystallised grains (therefore, non-optimised final microstructure) in the case of thicker final gauges (12 mm, for example) the same temperature may not be high enough to guarantee the complete recrystallisation of the as-cast grains. In both situations, the different total amount of reduction applied during rolling has completely modified the microstructural evolution.

To take into account these interactions, the evolution during rolling of several specific microstructural parameters has been recently evaluated with the help of the aforementioned model [235]. The parameters selected are the following:

–   Mean austenite grain size (D_{mean}).

–   Critical grain size (D_c): value for which at least a 10% of the volume fraction of grains have a larger size than that of the critical grain size.

–   Maximum grain size (D_{max}).

–   Microstructural heterogeneity (ZD): relationship between D_{max} and D_{mean}.

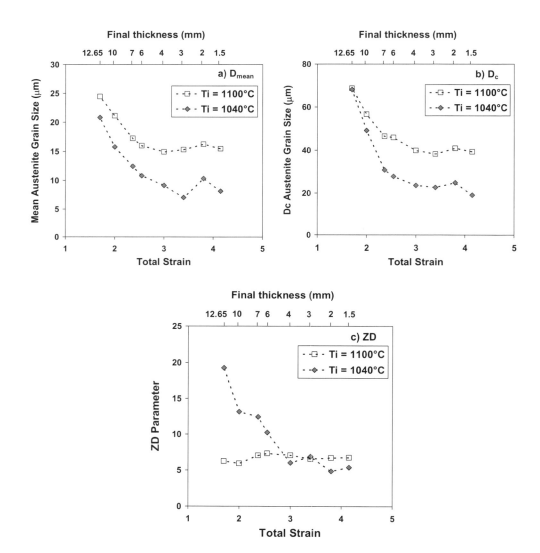

Fig. 88. Evolution of D_{mean}, D_c and ZD parameters during hot rolling as a function of final gauge thickness for the initial rolling temperatures of 1040 and 1100°C (0.06% C, 1.1% Mn, 0.2% Si, 0.035% Nb, 0.0080% N) [235].

The predictions obtained with the model are shown in Fig. 88 in the case of a 0.035% Nb microalloyed steel, as a function of the final gauge thickness. The simulations were made by selecting the rolling schedules listed in Table 16 and considering two initial rolling temperatures, 1040 and 1100°C, respectively.

If the evolution of mean grain size, D_{mean}, is considered, the predictions suggest that, with the initial rolling temperature of 1040°C, a finer microstructure could be achieved compared to the values obtained at 1100°C for all the range of final thicknesses considered. Similarly, Fig. 88a clearly illustrates the beneficial effect of applying larger total strains in microstructural refinement. It is also important to see the mean grain size evolution for gauge thicknesses larger than 6-7 mm. Nevertheless, with D_{mean} as the only parameter it is not possible to identify microstructural heterogeneities.

Table 16. Rolling parameters considered in computer simulations for final gauge thickness e (mm).

Pass	e = 1.5			e = 2			e = 3			e = 4			e = 6			e = 7			e = 10			e = 12.65			ΔT
	ε	$\dot{\varepsilon}$ (s⁻¹)	t_{ip} (s)	ε	$\dot{\varepsilon}$ (s⁻¹)	t_{ip} (s)	ε	$\dot{\varepsilon}$ (s⁻¹)	t_{ip} (s)	ε	$\dot{\varepsilon}$ (s⁻¹)	t_{ip} (s)	ε	$\dot{\varepsilon}$ (s⁻¹)	t_{ip} (s)	ε	$\dot{\varepsilon}$ (s⁻¹)	t_{ip} (s)	ε	$\dot{\varepsilon}$ (s⁻¹)	t_{ip} (s)	ε	$\dot{\varepsilon}$ (s⁻¹)	t_{ip} (s)	(°C)
1	1	5	6	0.9	5	6	0.75	5	6	0.7	5	6	0.55	5	6	0.5	5	6	0.5	5	6	0.4	5	6	35
2	1	15	3	0.9	15	3	0.75	10	4	0.7	10	4	0.55	10	4	0.5	10	4	0.5	10	4	0.4	10	4	30
3	0.85	50	1.8	0.8	40	2	0.7	15	3	0.55	15	3	0.55	15	3	0.45	15	3	0.45	15	5	0.4	15	5	30
4	0.55	90	1	0.5	70	1.6	0.5	40	2	0.45	30	2.1	0.4	25	2.5	0.4	25	2.5	—	—	—	—	—	—	30
5	0.4	150	0.7	0.4	90	1	0.4	60	1.7	0.35	50	1.8	0.35	40	2	0.3	30	2.1	0.3	20	2.7	0.3	20	2.7	30
6	0.3	200		0.3	120		0.3	80		0.25	70		0.25	60		0.25	50		0.25	25		0.2	25		(*)

(*) Dependent on the finishing rolling temperature at a constant cooling rate of 20°C/s up to 800°C.

If the D_c and ZD parameters are evaluated, it can be concluded that as the final gauge thickness increases there is a clear tendency to have a more pronounced microstructural heterogeneity. Concerning D_c, its evolution is very similar to that observed with D_{mean}. In contrast the ZD parameter provides additional information and clearly discriminates rolling conditions with similar D_{mean} and D_c values.

For example, in the case of the initial rolling temperature of 1040°C, for $e > 6$ mm, the high ZD values shown in Fig. 88c denote the presence of microstructural heterogeneities from the beginning of the rolling process that cannot be eliminated. In this case, the selection of a higher initial rolling temperature should contribute to the achievement of a more homogeneous final microstructure. Under these conditions, *ZD* takes values below 8, meaning that the degree of microstructural heterogeneity is relatively small.

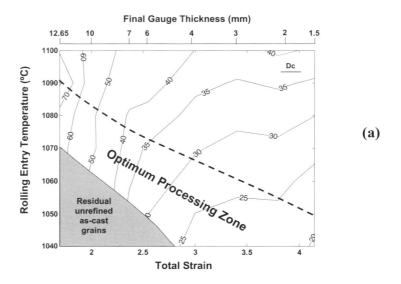

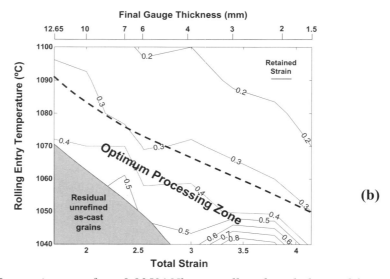

Fig. 89. Processing map for a 0.035% Nb microalloyed steel obtained from computer simulations [235]. Isoclines correspond to a) critical grain size D_c and b) retained strain.

With this type of analysis it is possible to build processing maps that can help in the selection of the best processing conditions for each specific final thickness and chemical composition [235]. An example of these maps is shown in Fig. 89 in the case of a 0.035% Nb steel and in Fig. 90 for a 0.05% Nb steel.

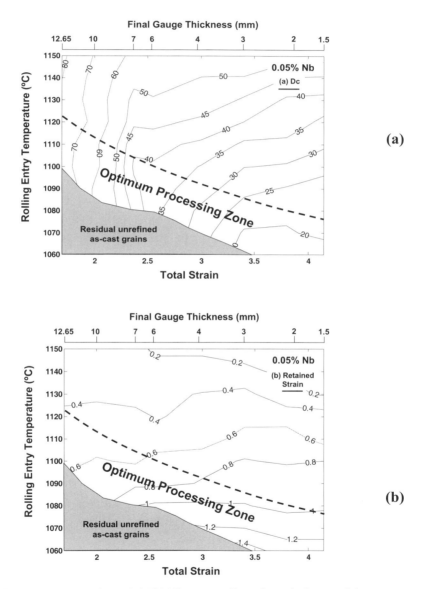

Fig. 90. Processing map for a 0.05% Nb microalloyed steel obtained from computer simulations [240]. Isoclines correspond to a) critical grain size D_c and b) retained strain.

In the maps depending on the initial rolling temperature and final gauge thickness, the processing conditions where residual as-cast grains remain in the final microstructure are defined. Inside these regions the model predicts that an as-cast grain volume fraction of 2%, at least, will remain in the final microstructure. Outside these regions the rest of the conditions plotted in the map give a value of the *ZD* parameter lower than 8, which can be considered as a reasonable degree of microstructural homogeneity. In the maps isoclines indicating the retained strain in austenite prior to transformation and the critical grain size D_c are drawn.

It is important to observe the relevance of Nb content on the size of the residual unrefined as-cast grains zone. As Nb content increases, higher initial rolling temperatures are recommended in order to favour complete recrystallisation of as-cast grains. On the other hand, higher Nb contents make it possible to increase the retained strain prior to transformation.

The combination of D_c isoclines (Fig. 89a and Fig. 90a) with retained strain isoclines (Fig. 89b and Fig. 90b) allows the optimum processing zone for each final gauge thickness to be defined. In the definition of this region, the following aspects have been considered:

–   The lack of the initial as-cast austenite.

–   No significant grain growth after final recrystallisation.

–   Convenience of retaining a certain amount of strain in order to enhance ferrite refinement during transformation.

From the optimum regions defined in the processing map, it turns out that, depending on the final gauge thickness, different initial rolling temperatures need to be selected.

In summary, with the help of the model described it is possible to define optimum TSDR processing conditions so as to achieve adequate levels of microstructural refinement and homogeneity.

This type of approach can be applied to different types of industrial plants based on TSDR technology taking into consideration the specific characteristics of each case:

–   Different initial thin slab thickness.

–   Different configurations of hot rolling mills, including number of stands, interpass times and possibility of cooling between passes.

5.4. OPTIMISATION OF ROLLING SCHEDULES

In the previous section the processing maps were determined considering the rolling schedules described in Table 16. These rolling schedules can be redesigned in order to optimise the austenite microstructure prior to transformation. This situation becomes more relevant in the case of thicker final gauges (for example $e > 7$-8 mm) combined with higher Nb contents. In the following, an example with a Nb content of 0.05% Nb is considered, starting out with a thin slab of 55 mm to obtain a final gauge of 10 mm.

The analysis of the microstructural evolution shows that the use of suitable combination of strain per pass / dummy passes may substantially improve the lack of initial unrecrystallised as-cast grains in the final product. To achieve this it is necessary to explore the possibilities of activating post-dynamic recrystallisation processes. This can be implemented by concentrating the applied strain in some of the rolling passes by introducing dummy passes in between.

The three examples selected are shown in Table 17 [240]. In the schedule *Seq 10* the five rolling passes listed previously in Table 16 are considered for comparison purposes. However in the modified sequences *Seq 10A* and *10B*, the same total strain is distributed into 4 passes, thus introducing two dummy passes and different combinations for the interpass times. An initial rolling temperature of 1070°C was selected, which, in agreement with the processing map in Fig. 90 for $e = 10$ mm and the *Seq* 10 rolling schedule, will lead to the presence of some fraction of unrecrystallised as-cast grains.

Table 17. Different rolling schedules corresponding to a final gauge thickness of 10 mm (initial thickness 55 mm).

Pass	Seq 10			Seq 10A			Seq 10B			
	ε	$\dot{\varepsilon}$ (s⁻¹)	t_{ip} (s)	ε	$\dot{\varepsilon}$ (s⁻¹)	t_{ip} (s)	ε	$\dot{\varepsilon}$ (s⁻¹)	t_{ip} (s)	ΔT (°C)
1	0.5	5	6	1	5	10	1	5	6	35
2	0.5	10	4	—	—	—	0.45	10	9	30
3	0.45	15	5	0.45	15	5	—	—	—	30
4	—	—	—	—	—	—	—	—	—	30
5	0.3	20	2.7	0.3	20	2.7	0.3	20	2.7	30
6	0.25	25		0.25	25		0.25	25		(*)

(*) Dependent on the finishing rolling temperature at a constant cooling rate of 20°C/s up to 800°C.

In Fig. 91 the microstructural evolution for the first three passes corresponding to Seq 10 is represented using the predicted austenite grain size distributions at the entry of each rolling pass [229]. Therefore, Fig. 91a would correspond to the initial as-cast grain size distribution. In the figure the different softening mechanisms intervening are indicated. In Range I the grains are under pure static recrystallisation conditions after the deformation pass, while in Range II the total strain is higher than ε_c, i.e. static and metadynamic recrystallisation mechanisms operate during the interstand time.

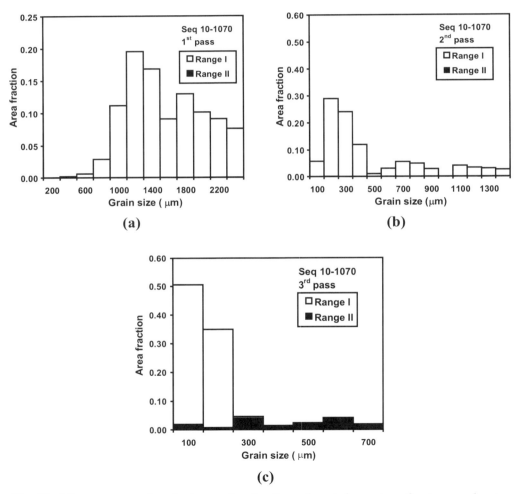

Fig. 91. *Microstructural evolution and activation of post-dynamic softening mechanisms for Seq 10 and* $T_i = 1070°C$. *Range I (*$\varepsilon < \varepsilon_c$*): pure static recrystallisation. Range II* ($\varepsilon_c < \varepsilon < \varepsilon_T$): *static + metadynamic recrystallisations [229].*

The activation of dynamic softening is delayed until the third pass, where an accumulated strain higher than ε_c is achieved in some of the grains (Fig. 91c). This strain is lower than the transition strain (ε_T) and as a consequence, both static and metadynamic processes will be activated during the interpass time. Nevertheless, the interpass time of 5 s after the third pass is not long enough to provide a complete refinement of the coarse grains and, prior to the 5th rolling pass, the microstructure still shows some heterogeneity (Fig. 92 [229]).

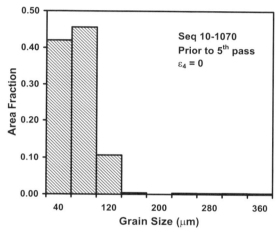

Fig. 92. Grain size distribution for Seq 10 just before the 5th rolling pass (at 925°C)
[229].

The microstructural evolutions calculated for the *Seq 10A* and *Seq 10B* are plotted in Fig. 93 [229]. A higher strain is applied in the first pass $(\varepsilon = 1$ vs. 0.5 for *Seq 10*) and different positions for dummy passes are defined (2nd and 4th passes for *Seq 10A* and 3rd and 4th for *Seq 10B*). For both *Seq 10A* and *Seq 10B*, the mechanisms acting after the 1st pass are the same and only one graph is plotted. The applied strain is higher than the critical strain for the onset of dynamic recrystallisation for grains smaller than 800 μm (Region II). For coarser grains, the model predicts that the static recrystallisation would be the only mechanism acting during the first interstand (Region I).

The microstructural distribution is slightly different at the entry of the following rolling pass (3rd in *Seq 10A* and 2nd in *Seq 10B*). As the interpass time is longer for *Seq 10A* a higher degree of recrystallisation is achieved, leading to a finer grain size distribution than for *Seq 10B*. Once the second deformation is applied, the coarsest grains for both sequences would reach a strain higher than the critical value, due to contribution of the retained strain coming from the first stand.

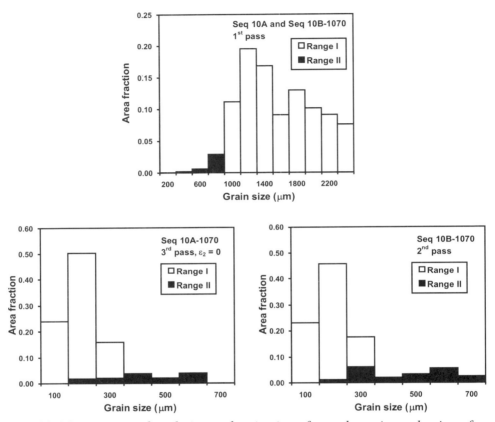

Fig. 93. Microstructural evolution and activation of post-dynamic mechanisms for Seq 10A and Seq 10B [229].

As a consequence the grain size distribution just before the next pass varies from *Seq 10A* to *Seq 10B*, as can be seen in Fig. 94. Due to higher temperature (1015°C for *Seq 10B* vs. 985°C for *Seq 10A*) and longer interpass times (9 s vs. 5 s), the evolution of post-dynamic recrystallisation in *Seq 10B* leads to a complete refinement of the initial as-cast structure. Meanwhile, during *Seq 10A*, a significant fraction of coarse grains remain unrecrystallised at the entry of the 5th pass and this could be the origin of microstructural heterogeneities in the final product.

When the grain size distributions of Fig. 92 and Fig. 94 are compared, the possibilities of reducing isolated microstructural heterogeneities through optimising the rolling schedules are clearly evident. Nevertheless, if high toughness requirements are necessary at very low temperatures, an additional grain size refinement should be mandatory. In this situation an increase in the total applied strain should be recommended. This implies that it is convenient to increase the thickness of initial thin slab, as has been pointed out by several authors [241, 242].

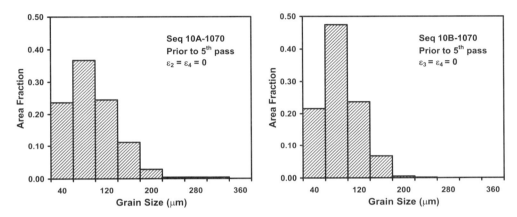

Fig. 94. Grain size distribution for Seq 10A and 10B at the entry of the 5th rolling pass (925°C) [229].

5.3 PHASE TRANSFORMATION

Regarding transformation in TSDR the general rules that are valid for cold charging can be applied. Nevertheless, in the case of TSDR processing some specific features related to microalloying with Nb and V should be mentioned.

Taking into account the characteristics of TSDR a significant amount of Nb may not precipitate during rolling, which means that Nb could remain in solid solution after rolling at concentrations significantly higher than those usually present in conventional CCR processes. This solute Nb may have different effects on the transformation products. An increase in hardenability is expected, leading to the formation of non-polygonal ferrite microstructures [238] (see example in Fig. 95). Furthermore, at certain conditions precipitation could occur at the same time as transformation may also take place. Both mechanisms can produce an important additional strengthening of the final microstructure compared to the levels achieved by simple ferrite grain size refinement mechanisms.

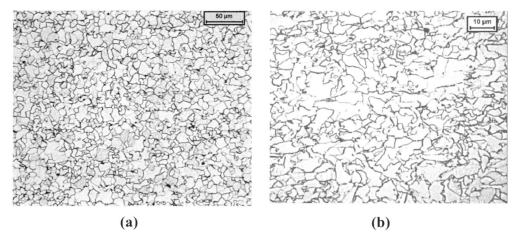

(a)   (b)

Fig. 95. Comparison between a) polygonal and b) non-polygonal ferrites.

There is a significant number of studies that mention the presence of very fine particles of MnS and CuS at different stages of TSDR [220, 221]. Independently of their influence on the evolution of the austenite microstructure during hot working, these particles can also act as particle-stimulated nucleation sites for ferrite grains. As mentioned previously in Chapter 3, there are different studies confirming that both MnS and CuS become active nucleation sites in V microalloyed steels, when V(C,N) coprecipitation takes place on the aforementioned particles. The presence of finer and more homogeneously distributed sulphide inclusions in TSDR conditions will contribute to an enhancement of ferrite grain size refinement. This beneficial effect has been quantified through an increase in the D_γ/D_α relationship in V-N microalloyed TSDR grades, compared to plain C-Mn steels [243].

6. INDUSTRIAL APPLICATIONS

The application of thermomechanical processings and microalloying has contributed significantly to an increase in TSDR steel production through the development of new and more complex steel grades. Because of the continuous worldwide spread of TSDR plants this trend in growth is expected to be maintain.

At the beginning the production in many TSDR plants was limited to hot strip of soft unalloyed steel grades [3, 182, 244]. In some cases, the steel had to meet cold rolled requirements as a coated material. In this sense, the reduction in free nitrogen content to improve deformability was necessary. For some specific qualities, this is achieved by adding boron, in other cases Ti microaddition is used [245]. In the case of scrap based EAF plants, in the raw material the reduction in the total nitrogen is achieved by combining scrap with different proportions of DRI products [206]. The initial production of low carbon steels was then followed by medium and high carbon grades. In this context, the development of very thin hot rolled strips, 0.8 mm, must be emphasised [246].

The following step was the development and production of HSLA steels. Nowadays, in several plants supplying to different markets, about 20-25% of the total production corresponds to HSLA steels [245]. In the following sections, some industrial developments concerning different steel grades will be discussed.

6.1. STRUCTURAL AND HSLA STEELS

One of the first microalloying applications to TSDR technology has been the development of different high strength hot strip grades. These qualities range from S315MC to S600MC. One of the main characteristics is the preference for low carbon contents (<0.07% C) in order to improve toughness and weldability [247]. While for the lower strength levels ferrite-pearlite microstructures work well, as requirements increase bainite contribution is necessary.

In general, for a specific steel grade, there is no single microalloying option. For example, if strength is considered as the main property, different approaches, based on V, Nb or both elements can be used. Usually, for structural steels up to 500 MPa yield stress, one element (Nb or V) is enough [3, 208, 243, 245], however, when higher strengths are required the combination of two microalloying elements is a better choice. The selection of one or another microalloying also means the definition of appropriate rolling schedules, as mention in Chapter 5. Fig. 96 shows the microstructure achieved by combining Nb and V in the case of a S500MC grade.

Another possibility is to combine one of these elements with Mo. In the case of V-Mo combinations high nitrogen contents (190-220 ppm) are selected to enhance precipitation strengthening [239].

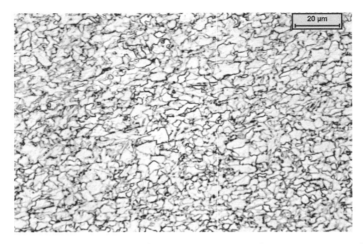

Fig. 96. Microstructure corresponding to a HSLA steel S500MC microalloyed with Nb and V.

The selection of one of the aforementioned elements can be determined by other factors, such as the characteristics of the steel (scrap based steel with high N content [153, 243] or BOF based steel with low N and residual elements), economical reasons or the metallurgical "know how" of the technicians involved in production.

6.2. DEVELOPMENT OF API GRADES

The implementation of API grades for line pipe applications in TSDR plants is one of the main challenges that has come about in the last few years, in a range which covers the interval from X52 to X80 [248]. It must be taken into account the fact that as API X number grade increases more complex mechanical requirements than those corresponding to structural steels are required. Usually, the residual element concentration (as S and P) should be reduced with a suitable secondary metallurgy practice. Another general characteristic is the selection of low carbon contents in order to obtain good performance and avoid risks of central segregation in the final product [249] which can induce delamination phenomenon on impact fracture specimens [250].

While in the lower grades (X52 and X60, for example [191]), single microadditions could be enough for mechanical purposes as the required strength level increases multiple additions, including combinations with Mo, are necessary [224]. As happens with structural steels, different approaches have been done for lower grades. For example, in the case of X52 grades an addition of 0.04% Nb (with 35 ppm nitrogen) is reported by El-Bitar et al. [191]. For the same grade, Xu et al. followed the approach of microalloying with vanadium (up to 0.06%), starting with 50 mm thin slab and rolling to final thicknesses of 7-8 mm [251].

Similarly, Megahed et al. selected the Nb-Ti combination to achieve the X60 quality with a final gauge thickness between 6 to 10 mm (rolling began with a thickness of 70 mm) [252]. In the case of X70, they increased Nb content up to 0.050% and added vanadium (0.04-0.05%).

As microalloying addition increases and thicker final gauges are required, there is a clear need to perform a very fine control of the thermomechanical parameters, following the main aspects considered in Chapter 5. This lack of optimisation may create a risk of microstructural heterogeneities appearing, as reported in [253] for the case of X56 grade.

Concerning toughness requirements, in agreement with the results published by Reip et al., to attain transition temperatures of -50 to -60°C slab/hot strip thickness ratios of around 7 to 8 are required [254]. These authors propose that for very extremely low temperature toughness requirements, for the case of 10 mm thick hot strip the thin slab thickness should be at least 80 mm. This again confirms the convenience of moving towards thicknesses that are greater than 50-60 mm for some specific applications.

6.3. DUAL PHASE STEELS

Dual phase and TRIP steels are becoming progressively more important in the automotive sector [255, 256]. This has increased their demand and TSDR has been considered as a valuable route to produce multiphase grades. One of the main advantages of dual phase steels is their good strength-formability combinations. This good performance is achieved through a proper ferritic matrix (70-95%) with embedded islands of martensite. The ferritic matrix provides high ductility while the martensitic phase contributes significantly to strength.

The production of hot rolled dual phase steels via TSDR requires some adjustments compared to the conventional route [257]. Once the strip leaves the last rolling stand (final rolling temperatures in the range of 800-850°C have been reported [258]), the steel is cooled down to the ferritic region (between

680 and 705°C [245]) and maintained so as to obtain the required ferrite fraction, before cooling rapidly to transform the remaining austenite (enriched by carbon) into martensite. In order to make the transformation easier, usually DP grades have different contents of Si and Cr.

Taking into account the fact that in typical CSP plants the length of the laminar cooling section is shorter than in conventional plants, to get the required ferrite fraction several modifications in the chemical composition and cooling conditions need to be carried out [258]. By doing this, DP 600 grade steel has successfully been produced. An example of microstructure corresponding to DP 600 is shown in the micrograph of Fig. 97.

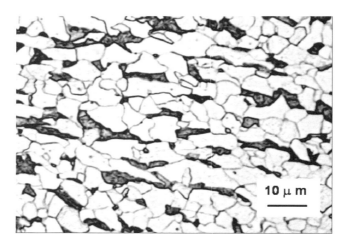

Fig. 97. Example of microstructure corresponding to TSDR DP 600 steel.

Finally, microalloying with Nb has been shown as a good procedure to refine the ferrite grain size in DP steels. Moreover, Nb makes the formation of the required fraction of ferrite during cooling easier [259]. This aspect becomes very relevant in TSDR technology because of the characteristics of run-out table (shorter than in conventional systems).

6.4.   CONCLUDING REMARKS

Several years ago thin slab technology became the most promising new development for the low cost production of microalloyed high strength steel strip [260].

The systematic studies carried out over the last ten years, taking into account the metallurgical and process peculiarities involved in TSDR, have allowed this technology to reach a high degree of maturity. In this context microalloying has become a key factor and one that it is widely exploited to achieve the more and more complex requirements of customers.

As has occurred in the cases mentioned above (structural, dual phase and pipe steels), in the near future the microalloying developments that are now being introduced in conventional production will be also implemented, adopting the necessary tailored solutions, in TSDR. In this context, thermomechanical processing of near-net-shape technologies, as it is the case of thin slab, can be considered as a fruitful field for future research and development and a way to maintain the leading role that steel has in modern world.

7. REFERENCES

[1]   J.P. Birat, R. Steffen and S. Wilmotte: State of the Art and Developments in Near-Net-Shape Casting of Flat Steel Products, Technical Steel Research, EUR 16671, (1995).

[2]   M. Hara, H. Kikuchi, M. Hanao, M. Kawamoto, T. Murakami and T. Watanabe: Rev. Metall. April (2002), p. 367.

[3]   P.J. Lubensky, S.L Wigman and D.J. Jonson: Microalloying'95 (ISS, Pittsburgh 1995), p. 225.

[4]   J. Watzinger, K. Moerwald and A. Wagner: Thinner Slab Casting. 33rd McMaster Symp. on Iron and Steelmaking (McMaster Univ., Hamilton 2005), p. 6.

[5]   F. Stella, P. Bobig, A. Carboni and I. Faruk: Intern. Symp. on Thin Slab and Rolling (The Chinese Society for Metals, Guangzhou 2002), p. 49

[6]   M. Millet: Thinner Slab Casting. 33rd McMaster Symp. on Iron and Steelmaking (McMaster Univ., Hamilton 2005), p. 1.

[7]   C. Yost: Commercialization of New Manufacturing Processes for Materials, Publication 3100 (U.S. International Trade Commission Office of Industries, USA 1998), p. 48.

[8]   R.J. Fruehan, O. Fortini, H.W. Paxton and R. Brindle: Theoretical Minimum Energies to Produce Steel (U.S. Dept. of Energy, Office of Industrial Technologies, Washington 2000).

[9]   N. Zapuskalov: ISIJ Inter. Vol. 43 (2003), p. 1115.

[10] J.M. Rodriguez-Ibabe: Mater. Sci. Forum Vol. 500-501 (2005) p. 49.

[11] T.N. Baker: Yield, Flow and Fracture of Polycrystals (Applied Science, London, 1983), p. 235.

[12] T. Gladman, D. Dulieu and I.D. McIvor: Microalloying'75 (Union Carbide, New York 1975), p. 32.

[13] B. Mintz, W.B. Morrison and A. Jones: Met. Technol. Vol. 6 (1979), p. 252.

[14] B. Mintz, G. Peterson and A. Nassar: Ironmaking Steelmaking Vol. 21 (1994), p. 215.

[15] P.D. Hodgson and R.K. Gibbs: ISIJ Int. Vol. 32 (1992), p. 1329.

[16] M.F. Ashby: Acta Met. Vol. 14 (1966), p. 679.

[17] T. Siwecki, A. Sandberg, W. Roberts and R. Langneborg: Thermomechanical Processing of Microalloyed Austenite (AIME, Pittsburgh 1982), 1653.

[18] R. Lagneborg, T. Siwecki, S. Zajac and B. Hutchinson: Scand. J. Metall. Vol. 28 (1999), p. 186.

[19] I.H.M. Ali, I.M. Moustafa, A. M. Farid and R.J. Glodowski: Mater. Sci. Forum Vol. 500-501 (2005), p. 503.

[20] A.J. DeArdo: Intern. Mater. Reviews Vol. 48 (2003), p. 371.

[21] F.B. Pickering and T. Gladman: ISI Special Report 81 (1963), p. 10.

[22] I.D. McIvor and T. Gladman: Research Report PMC/6805/1/75/a (British Steel, 1975).

[23] G.R. Speich and H. Watlimont: JISI Vol. 206 (1968), p. 385.

[24] F.B. Pickering: Microalloying´75 (Union Carbide, New York 1975), p. 9.

[25] T. Gladman: The Physical Metallurgy of Microalloyed Steels (IoM, London 1997), p. 64

[26] W.B. Morrison, R.C. Cochrane and P.S. Mitchell: ISIJ Int. Vol. 33 (1993), p. 1095.

[27] C. J. McMahon and M. Cohen: Acta Metall. Vol. 13 (1965), p. 591.

[28] J.F. Knott: Reliability and Structural Integrity of Advanced Materials, ECF9, Vol. 2 (EMAS, Varna 1992), p. 1375

[29] J. H. Tweed and J. F. Knott: Acta Metall. Vol. 35 (1987), p. 1401.

[30] T. Lin, A. G. Evans and R. O. Ritchie: Met. Trans. A Vol. 18A (1987), p. 641.

[31] M.A. Linaza, J.L. Romero, J.M. Rodriguez-Ibabe and J.J. Urcola: Scr. Metall. Mater. Vol. 29 (1993), p. 451.

[32] P. Bowen, M.B.D. Ellis, M. Strangwood and J.F. Knott: Fracture Control of Engineering Structures, ECF6, Vol. 3 (EMAS, Amsterdam 1986), p. 1751.

[33] D.E. McRobie and J.F. Knott: Mater. Sci. Technol. Vol. 1 (1985), p. 357.

[34] I. San Martin: PhD Thesis (Univ. Navarra, San Sebastian 1998)

[35] M.A. Linaza, J.L. Romero, J.M. Rodriguez-Ibabe and J.J. Urcola: Scr. Metall. Mater. Vol. 29 (1993), p. 1217.

[36] A. Echeverria: unpublished work (CEIT).

[37] D. Brooksbank and K.W. Andrews: JISI Vol. 206 (1968), p. 595.

[38] A.R. Rosenfield and D.K. Shetty: ASTM STP 856 (1985), p. 196.

[39] M.A. Linaza, J.L. Romero, J.M. Rodriguez-Ibabe and J.J. Urcola: 36th Mechanical Working and Steel Proc. Conf. Vol. 32, (ISS, Baltimore 1995), p. 483.

[40] T.J. Baker, F.P.L. Kavishe and J. Wilson: Mater. Sci. Technol. Vol. 2 (1986), p. 576.

[41] J.L. Lewandowsky and A.W. Thompson: Metall. Trans. A Vol. 17A (1986), p. 1769.

[42] A. Echeverria, M.A. Linaza and J.M. Rodriguez-Ibabe: Mater. Sci. Forum Vol. 284-286 (1998), p. 351.

[43] M.A. Linaza, J.L. Romero, I. San Martin, J.M. Rodriguez-Ibabe and J.J. Urcola: Microalloyed Bar and Forging Steels (TMS, Golden 1996), p. 311.

[44] M.A. Linaza, J.L. Romero, J.M. Rodriguez-Ibabe and J.J. Urcola: Scr. Metall. Mater. Vol. 32 (1995), p. 395.

[45] M.A. Linaza, J.M. Rodriguez-Ibabe and J.J. Urcola: Fatigue Fract. Engng. Mater. Struct. Vol. 20 (1997), p. 619.

[46] I. San Martin and J.M. Rodriguez-Ibabe: Scr. Mater. Vol. 40 (1999), p. 459.

[47] J.M. Hyzak and I.M. Bernstein: Metall. Trans. A Vol. 7A (1976), p. 1217.

[48] D.J. Alexander and I.M. Bernstein: Metall. Trans. A Vol. 13A (1982), p. 1865.

[49] E. Cotrina, B. Lopez and J.M. Rodriguez-Ibabe: Austenite Formation and Decomposition (TMS, Chicago 2003), p. 213.

[50] J.P. Naylor and P.R. Krahe: Metall. Trans. A Vol. 5A (1974), p. 1699.

[51] P. Brozzo, G. Buzzichelli, A. Mascanzoni and M. Mirabile: Met. Sci. (1977), p. 123.

[52] M. Diaz, I. Madariaga, J.M. Rodriguez-Ibabe and I. Gutierrez: J. Construct. Steel Res. Vol. 46 (1998), p. 413.

[53] M. Diaz-Fuentes, A. Iza-Mendia and I. Gutierrez: Metall. Mater. Trans. A Vol. 34A (2003), p. 2505.

[54] D. Bhattacharjee, J.F. Knott and C.L. Davis: Metall. Mater. Trans. A Vol. 35A (2004), p. 121.

[55] A. Echeverria and J.M. Rodriguez-Ibabe: Mater. Sci. Eng. Vol. 346A (2003), p. 149.

[56] A.I. Fernández, P. Uranga, B. López and J.M. Rodriguez-Ibabe: Fracture Mechanics. Applications and Challenges ECF 13, (Elsevier, San Sebastian 2000), CD-ROM.

[57] T. Sakai and J.J. Jonas: Acta Metall. Vol. 32 (1984), p. 189.

[58] J.P. Sah, G.J. Richardson and C.M. Sellars: Met. Sci. Vol. 8 (1974), p. 325.

[59] C.M. Sellars: Hot Working and Forming Processes (Metals Soc., London 1980), p. 3.

[60] C.M. Sellars: Mater. Sci. Technol. Vol. 6 (1990), p. 1072.

[61] W.P. Sun and E.B. Hawbolt: ISIJ Int. Vol. 37 (1997), p. 1000.

[62] S.F. Medina and C.A. Hernández: Acta Metall. Vol. 44 (1996), p. 149.

[63] C. Roucoules, S. Yue and J.J. Jonas: Proc. Int. Conf. on Modeling of Metal Rolling Processes (IoM, London 1993), p. 165.

[64] A.I. Fernández, B. López and J.M. Rodriguez-Ibabe: Mater. Sci. Forum Vol. 467-470 (2004), p. 1169.

[65] A.I. Fernández, P. Uranga, B. López and J.M. Rodriguez-Ibabe: Mater. Sci. Eng. Vol. 361A (2003), p. 367.

[66] J.J. Jonas: High Strength Low Alloy Steels (Wollongong 1984), p. 80.

[67] K. Minami, F. Siciliano, T.M. Maccagno and J.J. Jonas: ISIJ Int. Vol. 36 (1996), p. 1507.

[68] M.G. Akben and J.J. Jonas: HSLA Steels. Technology and Application (ASM, Ohio 1983), p. 149.

[69] B. Pereda: unpublished work (CEIT).

[70] F. Siciliano and J.J. Jonas: Metall. Mater. Trans. A Vol. 31A (2000), p. 511.

[71] L.E. Cepeda, J.M. Rodriguez-Ibabe and J.J. Urcola: Mater. Sci. Forum Vol. 113-115 (1993), p. 399.

[72] D.Q. Bai, S. Yue and J.J. Jonas: Thermomechanical Processing of Steel (Metals Society, Montreal 2000), p. 669.

[73]   P. Uranga, A.I. Fernández, B. López and J.M. Rodriguez-Ibabe: Mater. Sci. Eng. Vol. 345A (2003), p. 319.

[74]   A.I. Fernández, P. Uranga, B. López and J.M. Rodriguez-Ibabe: ISIJ Int. Vol. 40 (2000), p. 893.

[75]   P. Patel, C. Zhou and R. Priestner: 3rd Int. Conf. on Recrystallization and Related Phenomena, Rex'96, (Monterrey, 1996), p. 421

[76]   A. Laasraoui and J. J. Jonas: Metall. Trans. A Vol. 22A (1991), p. 151.

[77]   M. Arribas, B. López and J.M. Rodriguez-Ibabe: Mater. Sci. Forum Vol. 500-501 (2005), p. 131.

[78]   S. F. Medina and J. E. Mancilla: ISIJ Int. Vol. 36 (1996), p. 1070

[79]   E. Ruibal, J.J. Urcola and M. Fuentes: Z. Metallkde Vol. 76 (1985), p. 568.

[80]   K. Airaksinen, L.P. Karjalainen, D. Porter and J. Perttula: Mater. Sci. Forum Vol. 284-286 (1998), p. 119.

[81]   S.F. Medina, J.E. Mancilla and C.A. Hernandez: ISIJ Int. Vol. 34 (1994), p. 689.

[82]   R. Priestner and C. Zhou: Ironmaking Steelmaking Vol. 22 (1995), p. 326.

[83]   C. García-Mateo, B. López and J.M. Rodriguez-Ibabe: Mater. Sci. Eng. Vol. A303 (2001), p. 216.

[84]   C. Roucoules, P.D. Hodgson, S. Yue and J.J. Jonas: Metall. Trans. A Vol. 25A (1994), p. 389.

[85]   P.D. Hodgson, L.O. Hazelden, D.L. Matthews and R.E. Gloss: Microalloying'95 (ISS, Pittsburgh 1995), p. 341.

[86]   W. Roberts, A. Sandberg, T. Siwecki and T. Werlefors: Conf. on HSLA Steels Technology and Applications (ASM, Philadelphia 1984), p. 67.

[87]   S-H. Cho, K-B. Kang and J.J. Jonas: ISIJ Int. Vol. 41 (2001), p. 63.

[88]   J.J. Jonas: Mater. Sci. Eng. Vol. 184A (1994), p. 155.

[89]   J.J. Jonas: Mater Sci. Forum Vol. 284-286 (1998), p. 3.

[90]   S. Yue, C. Roucoules, T.M. Maccagno and J.J. Jonas: 37th Mechanical Working and Steel Proc. Conf. Vol. 33 (ISS, Hamilton 1996), p. 651.

[91]   C. Roucoules, S. Yue and J.J. Jonas: Metall. Mater. Trans. A Vol. 26A (1995), p. 181.

[92]   J.H. Beynon and C.M. Sellars: ISIJ Int. Vol. 32 (1992), p. 359.

[93]   R. Abad, A.I. Fernández, B. López and J.M. Rodriguez-Ibabe: ISIJ Int. Vol. 41 (2001), p. 1373

[94]   C. Devadas, I.V. Samarasekeera and E.B. Hawbolt: Metall. Trans. A Vol. 22A (1991), p. 335.

[95]   C.M. Sellars and J.A. Whiteman: Met. Sci. Vol. 13 (1979), p. 187.

[96]   I. Tamura: Thermec'88 (ISIJ, Tokyo 1988), p. 1.

[97]   T. Tanaka: Microalloying'95 (ISS, Pittsburg 1995), p. 165.

[98]   R. Bengochea, B. López and I. Gutierrez: ISIJ Int. Vol. 39 (1999), p. 583.

[99]   M.G. Akben, I. Weiss and J.J. Jonas: Acta Metall. Vol. 29 (1981), p. 111.

[100] W.P. Sun, W.J. Liu and J.J. Jonas: Metall. Trans. A Vol. 20A (1989), p. 2707.

[101] O. Kwon and A.J. DeArdo: Acta Met. Vol. 39 (1991), p. 529.

[102] K. Narita: Trans. ISIJ Vol. 15 (1975), p. 145

[103] W. Leslie, R.L. Rickett, C.L. Dotson and C.S. Walton: Trans. ASM Vol. 46 (1954), p. 1470.

[104] E.T. Turkdogan: Fundamentals of Steelmaking (IoM, London 1996).

[105] K.J. Irvine, F.B. Pickering and Y. Gladman: JISI Vol. 205 (1967), p. 161.

[106] S. Matsuda and N. Okumura: Trans. ISIJ Vol. 18 (1978), p. 198.

[107] S.S. Hansen, J.B. Vander Sande and M. Cohen: Metall. Trans. A Vol. 11A (1980), p. 387.

[108] E.J. Palmiere, C.I. Garcia and A.J. DeArdo: Met. Trans. A Vol. 27A (1996), p. 951.

[109] H.J. Frost and M.F. Ashby: Deformation Mechanisms Maps (Pergamon Press, Oxford 1982).

[110] M. Gomez and S.F. Medina: Mater. Sci. Forum Vol. 500-501 (2005), p. 147.

[111] B. Dutta and C.M. Sellars: Mater. Sci. Technol. Vol. 22 (1987), p. 1511.

[112] D.Q. Bai, S. Yue, W.P. Sun and J.J. Jonas: Metall. Trans. A Vol. 24A (1993), p. 2151.

[113] L.P. Karjalainen, T.M. Maccagno and J.J. Jonas: ISIJ Int. Vol. 35 (1995), p. 1523.

[114] R. Abad, B. López and I. Gutierrez: Mater. Sci. Forum Vol. 284-286 (1998), p. 167.

[115] R. Abad, B. López and J.M. Rodriguez-Ibabe: J. Mater Proc. Technol. Vol. 117/3 (2001), CD-ROM Section C4.

[116] P. Choquet, A. Le Bon and Ch. Perdrix: Proc. Int. Conference on the Strength of Metals and Alloys, ICSMA 7 (Pergamon Press, Montreal 1985) Vol. 2, p. 1025.

[117] Ch. Perdrix: Characteristic of Plastic Deformation of Metals During Hot Working, ECSC Report, No. 7210 EA/31, IRSID, 1987.

[118] J.H. Beynon, A.R.S. Ponter and C.M. Sellars: Proc. Int. Conf. Modelling of Metall Forming Processes, (Kluwer Academic Publ., 1988), p. 321.

[119] X. Liu, J.K. Solberg, R. Gjengedal and A.O. Kluken: Mater. Sci. Technol. Vol. 11 (1995), p. 469.

[120] C. Devadas, I.V. Samarasekera and E. B. Hawbolt: Metall. Trans. A Vol. 22A (1991), p. 335.

[121] P. L. Orsetti and C.M. Sellars: Acta Mater. Vol. 45 (1997), p. 137.

[122] A. Fernández, B. López and J.M. Rodriguez-Ibabe: Scr. Mater. Vol. 46 (2002), p. 823.

[123] E. Anelli: ISIJ Int. Vol. 32 (1992), p. 440.

[124] I. Gutierrez, B. López and J.M. Rodriguez-Ibabe: First Joint Intern. Conf. on Recrystallization and Grain Growth (Springer-Verlag, Aachen 2001), p. 145.

[125] F.B. Pickering: Ti Technology in Microalloyed Steels (IoM, 1997), p. 10.

[126] S. Zajac, T. Siwecki, B. Hutchinson and M. Attlegard: Metall. Trans. A Vol. 22A (1991), p. 2681.

[127] Y. Misaka and T. Yoshimoto: J. Jpn. Soc. Technol. Plast. Vol. 8 (1967), p. 414.

[128] S. Shida: Hitachi Res. Lab. Report (1974), p. 1.

[129] T.M. Maccagno, J.J. Jonas, S. Yue, B.J. McCrady, R. Slobodian and D. Deeks: ISIJ Int. Vol. 34 (1994), p. 917.

[130] J.G. Lenard, M. Pietrzyk and L. Cser: Mathematical and physical simulation of the properties of hot rolled products, Elsevier (1999), p. 90.

[131] R. Bengochea, B. Lopez and I. Gutierrez: Mater. Sci. Forum Vol. 284-286 (1998), p. 201.

[132] I. Kozasu, C. Ouchi, T. Sampei and T. Okita: Microalloying'75 (Union Carbide, New York 1975), p. 120.

[133] R. Bengochea: CEIT internal Report (1998).

[134] A. Sandberg and W. Roberts: Intern. Conf. on Thermomechanical Processing of Microalloyed Austenite (AIME, Pittsburgh 1981), p. 405.

[135] J.H. Beynon and C.M. Sellars: High Strength Low Alloy Steels (Wollongong, 1984), p. 184.

[136] R.K. Gibbs, B.A. Parker and P.D. Hodgson: Intern. Symp. on Low-Carbon Steels for the 90`s (TMS, Warrendale 1993), p. 173.

[137] R. Zubialde, B. López and J.M. Rodriguez-Ibabe: Mater. Sci. Forum Vol. 500-501 (2005), p. 403.

[138] B. Donnay, J.C. Herman, V. Leroy, U. Lotter, R. Grossterlinden and H. Pircher: Modelling of Steel Microstructural Evolution During Thermomechanical Treatment, EUR 17585 (European Commission, Luxemburg 1997), p. 113.

[139] E. Cotrina, A. Iza-Mendia and B. Lopez: unpublished work (CEIT).

[140] S. Zajac: Mater. Sci. Forum Vol. 500-501 (2005), p. 75.

[141] C. Klinkenberg, K. Hulka and W. Bleck: Steel Research Int. Vol. 75 (2004), p. 744.

[142] A. Altuna and I. Gutierrez: unpublished work (CEIT).

[143] T. Kimura, A. Ohmori, F. Kawabata and K. Amano: Thermec'97 (TMS, Warrendale 1997), p. 645.

[144] T. Kimura, F. Kawabata, K. Amano, A. Ohmori, M. Okatsu and K. Uchida: Intern. Symp. on Steel for Fabricated Structures (ASM, Cincinnati 1999), p. 165.

[145] S. Zajac: 43rd Mechanical Working and Steel Proc. Conf. (ISS, Charlotte 2001), p. 497.

[146] T. Furuhara and T. Maki: Mater. Sci. Eng. A Vol. 312 (2001), p. 145.

[147] Z. Guo, N. Kimura, S. Tagashira, T. Furuhara and T. Maki: ISIJ Int. Vol. 42 (2002) p. 1033

[148] D. Hernández, M. Díaz-Fuentes, B. López and J.M. Rodriguez-Ibabe: Mater. Sci. Forum Vol. 426-432 (2003), p. 1151.

[149] D. Hernández, B. López and J.M. Rodriguez-Ibabe: Mater. Sci. Forum Vol. 500-501 (2005), p. 411.

[150] D. Hernández, B. López and J.M. Rodriguez-Ibabe: Conf. Proc. Microalloyed Steels 2002 (ASM Intern., Columbus 2002), p. 64.

[151] O. Kwon: Rev. Metall. January (2003), p. 25.

[152] K. Miyazawa: Science and Technol. of Advanced Materials Vol. 2 (2001) p. 59.

[153] M. Korchynsky and S. Zajac: Thermomechanical Processing in Theory, Modelling and Practice, (ASM Intern., Stockholm 1996), p. 369.

[154] N. Zapuskalov: ISIJ Int. Vol. 43 (2003), p. 1115.

[155] R. Kaspar and O. Paweslki: METEC Congress 94 Vol. 1 (VDEh, Düsseldorf 1994), p. 390

[156] W. Löser, S. Thiem and M. Jurisch: Mater. Sci. Eng., Vol. A173 (1993), p. 323.

[157] X. Huo, D. Liu, Y. Wang, N. Chen, Y. Kang and J. Fu: Journ. Univers. Science and Technology Beijing Vol. 11 (2004), p. 133.

[158] H. Yu, Y. Kang, K. Wang, J. Fu, Z. Wang and D. Liu: Mater. Sci. Eng. Vol. 363A (2003), p. 86.

[159] Ruizhen Wang, C. I. Garcia, M. Hua, Hongtao Zhang and A.J. DeArdo: Mater. Sci. Forum Vol. 500-501 (2005), p. 229.

[160] Y-M. Won and B.G. Thomas: Metall. Mater. Trans. A Vol. 32A (2001), p. 1755.

[161] J.E. Camporredondo, A.H. Castillejos, F.A. Acosta, E.P. Gutierrez and M.A. Herrera: Metall. Mater. Trans. B Vol. 35B (2004), p. 541.

[162] G. Krauss: Metall. Mater. Trans. B Vol. 34B (2003), p. 781.

[163] J. Konishi, M. Militzer, J.K. Brimacombe and I.V. Samarasekera: Metall. Mater. Trans. B Vol. 33B (2992), p. 413.

[164] J.K. Brimacombe and I.V. Samarasekera: Turkdogan Symposium Proceedings, (ISS, Pittsburgh 1994), p. 171.

[165] V. Guyot, J-F. Martin, A. Ruelle, A. d'Anselme, J-P. Radot, M. Bobadilla, J-Y. Lamant and J-N. Pontoire: ISIJ Int. Vol. 36 (1996), p. S227.

[166] C. Bernhard and G. Xia: Ironmaking Steelmaking Vol. 33 (2006), p. 1.

[167] S.V. Subramanian and H. Zou: Proc. Intern. Conf. on Processing Microstructure and Properties of Microalloyed and Other Modern HSLA Steels (ISS, Pittsburgh 1991) p. 23.

[168] W.F. Gale and T.C. Totemeier: Smithells Metals Reference Book, (ASM Intern., 2004), p. 13-1

[169] S. Kurokawa, J.E. Ruzzante, A.M. Hey and F. Dymat: 36th Annual Cong. ABM Vol. 1 (Refice, Brazil 1981), p. 47.

[170] M.C.M. Cornelissen: Ironmaking Steelmaking Vol. 16 (1986), p. 204.

[171] X.M. Zhang, Z.Y. Jiang, X.H. Liu and G.D. Wang: J. Mater. Process. Technol. Vol. 162-163 (2005), p. 591.

[172] N. Yoshida, O. Umezawa and K. Nagai: ISIJ Int. Vol 43 (2003), p. 348.

[173] Z. Han, K. Cai and B. Liu: ISIJ Int. Vol. 41 (2001), p. 1473.

[174] K. Hulka and F. Heisterkamp: Mater. Sci. Forum Vol. 284-286 (1998), p. 343.

[175] P. Bordignon and K. Hulka: HSLA Steels 2005 and ISUGS 2005 (CSM, Sanya 2005), p. 45.

[176] T. Taira, K. Matsumoto, Y. Kobayashi, K. Takeshige and I. Kozasu: HSLA Steels. Technology and Application (ASM, Ohio 1983), p. 723.

[177] J.E. Camporredonde, F.A. Acosta, A.H. Castillejos, E.P. Gutierrez and R. Gonzalez: Metall. Mater. Trans. B Vol. 35B (2004), p. 561.

[178] S. Ogibayashi, M. Yamada, Y. Yoshida and T. Mukai: ISIJ Int. Vol. 31 (1991), p. 1408.

[179] C. Beckermann: Encyclopedia of Materials: Science and Technology, Elsevier (2001), p. 4733.

[180] J.S. Ha. J.R. Cho, B.Y. Lee and M.Y. Ha: J. Mater. Proc. Technol. Vol. 113 (2991), p. 257.

[181] U. Yoon, I-W. Bang, J.H. Rhee, S-Y. Kim, J-D. Lee and K.H. Oh: ISIJ Int. Vol. 42 (2002), p. 1103.

[182] C. Hendricks, W. Rasim, H. Janssen, H. Schnitzer, E. Sowka, P. Tesè: Rev. Metall. July (2001), p. 656.

[183] A. Yamanaka, S. Kumakura, K. Okamura, T. Kanazawa, T. Muramaki, M. Oka, I. Takeuchi and T. Watanabe: Ironmaking Steelmaking Vol. 26 (1999), p. 457.

[184] F.P. Pleschiutschnigg, B. Krüger, P. Meyer, G. Cosio, U. Siegers and H.G. Gusgen: Rev. Metall. June (1992), p. 547.

[185] M.D.C. Sobral, P.R. Mei, R.G. Santos, F.C. Gentile and J.C. Bellon: Ironmaking Steelmaking Vol. 30 (2003), p. 412.

[186] W.R. Irving: Continuous Casting of Steel (IoM, London 1993), p. 74.

[187] B. Mintz: ISIJ Int. Vol. 39 (1999), p. 833.

[188] K. Banks, A. Kourtsaris, F. Verdoorn and A. Tuling: Mater. Sci. Technol. Vol. 17 (2001), p. 1596.

[189] B. Mintz: Ironmaking Steelmaking Vol. 27 (2000), p. 343.

[190] B. Mintz, S. Yue and J.J. Jonas: Intern. Mater Rev. Vol. 36 (1991), p. 187.

[191] T. El-Bitar, K.-E. Hensger, C. Klinkenberg, T.E. Weirich, G. Megahed, A. El-Kady and G. Hefny: 2nd Intern. Conf. on Thermomechanical Processing of Steels (Verlag Stahleisen, Düsseldorf 2004), p. 187.

[192] J. Ruf, C. Davis, C. Miller and J. Weglo: Thinner Slab Casting. 33rd McMaster Symp. on Iron and Steelmaking (McMaster Univ., Hamilton 2005), p. 176.

[193] P. Uranga, A.I. Fernández, B. López and J.M. Rodriguez-Ibabe: 43rd Mechanical Working and Steel Proc. Conf. Vol. 39, (ISS, Charlotte 2001), p. 511.

[194] N.S. Pottore, C.I. Garcia and A.J. DeArdo: Metall Trans. A Vol. 22A (1991), p. 1871.

[195] L.D. Frawley and R. Priestner: Mater. Sci. Forum Vol. 284-286 (1998), p. 485.

[196] N. Yoshida, O. Umezawa and K. Nagai: ISIJ Int. Vol. 43 (2003), p. 348.

[197] N. Yoshida, O. Umezawa and K. Nagai: ISIJ Int. Vol. 44 (2004), p. 547.

[198] R. Wang, C.I. Garcia, M. Hua, H. Zhang and A.J. DeArdo: Mater. Sci. Forum Vols. 500-501 (2005), p. 229.

[199] M.P. Guerrero-Mata, A.L. Delgado, P.C. Zambrano, L.A. Leduc and R. Colás: Thermomechanical Processing: Mechanics, Microstructure and Control (Univ. of Sheffield, Sheffield 2002), p. 309.

[200] Z. Hongtao, P. Ganyun, W. Ruizhen and L. Chengbin: Mater. Sci. Forum Vols. 500-501 (2005), p. 295.

[201] L. Habraken and J. Lecomte-Beckers: Copper in Iron and Steel (J. Wiley & Sons 1982), p. 45

[202] M.J.U.T. van Wijngaarden and G.P. Visagie: Steelmaking Conf. Proc. Vol. 79 (ISS, Pittsburgh 1996), p. 627.

[203] N. Imai, N. Komatsubara and K. Kunishige: ISIJ Int. Vol. 37 (1997) p. 217.

[204] C. Kremer, B. Debyser and G. Schanne: Rev. Metall. Sept. (1991), p. 913.

[205] S.L. Wigman and M.D. Millett: Scaninject VI (Lulea 1992).

[206] F. Aristegui, J.I. Langara, I. Lasa, A. Fernandez, J. Miñambres and A. Calderon: Rev. Metall-CIT Vol. (2003), p. 637.

[207] G. Walmag, M. Picard and N. Mikler: Scale Control When Direct Rolling, Technical Steel Research EUR 21417 (European Commission, Luxembourg 2005).

[208] D.N. Crowther, Y. Li, T.N. Baker, M.J.W. Green and P.S. Mitchell: Thermomechanical Processing of Steels (IoM, London 2000), p. 527.

[209] Y. Li, D.N. Crowther, P.S. Mitchell and T.N. Baker: ISIJ Int. Vol. 42 (2002), p. 636.

[210] P.S. Mitchell: Thinner Slab Casting. 33rd McMaster Symp. on Iron and Steelmaking (McMaster Univ., Hamilton 2005), p. 277..

[211] J.S. Park, M. Ajmal and R. Priestner: ISIJ Int. Vol. 40 (2000), p. 380.

[212] S. Jacobs, B. Soenen and C. Klinkenberg: Intern. Symp. on Thin Slab Casting and Rolling (CSM, Guangzhou 2002), p. 368.

[213] B. Soenen, S. Jacobs and C. Klinkenberg: Microalloyed Steels (ASM Intern., Columbus 2002), p. 16.

[214] M.T. Nagata, J.G. Speer and D.K. Matlock: Metall. Mater. Trans. A Vol.33A (2002), p. 3099.

[215] R.F. Gibbs, R. Peterson and B.A. Parker: Proc. Int. Conf. on Processing, Microstructure and Properties of Microalloyed and Other Modern HSLA Steels (AIME, Warrendale 1991), p. 201.

[216] M. Arribas, B. López and J.M. Rodriguez-Ibabe: Intern. Conf. on New Developments in Long and Forged Products: Metallurgy and Applications (AIST, Winter Park 2006), p. 143.

[217] D. Hernández, Ph Thesis, University of Navarra, 2003.

[218] R. Kaspar, Steel Res. Vol. 74 (2003), p. 318.

[219] P. Uranga, A.I. Fernández, B. López and J.M. Rodriguez-Ibabe: Thermomechanical Processing of Steels (IoM, London 2000), p. 204.

[220] K. Kawasaki, T. Senuma, S. Akamatsu, T. Hayashida, S. Sanagi and O. Akisue: Thermec'97 (TMS, Warrendale 1997), p. 629.

[221] V. Leroy and J.C. Herman: Microalloying'95 (ISS, Pittsburgh 1995), p. 213.

[222] Y. Li, J.A. Wilson, D.N. Crowther, P.S. Mitchell, A.J. Craven and T.N. Baker: ISIJ Int. Vol. 44 (2004), p. 1093.

[223] T.N. Baker, Y. Li, J.A. Wilson, A.J. Craven and D.N. Cowther: Mater. Sci. Technol. Vol. 20 (2004), p. 720.

[224] C.I. Garcia, C. Torkaz, C. Graham and S.J. DeArdo: 2nd Intern. Conf. on Thermomechanical Processing of Steels (Verlag Stahleisen, Düsseldorf 2004), p. 173.

[225] R. Wang, M. Hua, H. Zhang and C.I. Garcia: 5th Intern. Conf on HSLA Steels (Sanya, China 2005), p. 292.

[226] P.S. Mitchell: Mater. Sci. Forum Vol. 500-501 (2005), p. 269.

[227] K.E. Hensger and G. Flemming: Niobium Science & Technology (TMS, Orlando 2001), p. 405.

[228] S. V. Subramanian, G. Zhu, H.S. Zurob, G.R. Purdy, G.C. Weatherly, J. Patel, C. Klinkenberg and R. Kaspar: Thermomechanical Processing: Mechanics, Microstructure and Control (Univ. of Sheffield, Sheffield 2003), p. 148.

[229] P. Uranga, A.I. Fernández, B. López and J.M. Rodriguez-Ibabe: Materials Science and Technology MST'06. Steel-related Papers (AIST, Cincinnati 2006), p. 781.

[230] A.I. Fernández, B. López and J.M. Rodriguez-Ibabe: Thermomechanical Processing: Mechanics, Microstructure and Control (Univ. of Sheffield, Sheffield 2002), p. 302.

[231] P. Uranga, A.I. Fernández, B. López and J.M. Rodriguez-Ibabe: ISIJ Int. Vol. 44 (2004), p. 1416.

[232] K. Matsuura and Y. Itoh: Materials Transactions JIM Vol. 32 (1991), p. 1042.

[233] M. Militzer, A. Giumelli, E. B. Hawbolt and T. R. Meadowcroft: Metall. Mater. Trans. A Vol. 27A (1996), p. 3399.

[234] A. I. Fernández, B. López and J. M. Rodriguez-Ibabe: Metall. Mat. Trans. A, Vol. 33A (2002), p. 3089.

[235] P. Uranga, A.I. Fernández, B. López and J.M. Rodriguez-Ibabe: Mater. Sci. Forum Vol. 500-501 (2005), p. 245.

[236] P. Uranga, A.I. Fernández, B. López and J.M. Rodriguez-Ibabe: Mater. Sci. Forum Vol. 426-432 (2003), p. 3915.

[237] J.M. Rodriguez-Ibabe, P. Uranga, A.I. Fernández, B. Aizpurua and B. López: Intern. Symp. on Thin Slab Casting and Rolling (CSM, Guangzhou 2002), p. 252

[238] A.J. DeArdo, M. Hua, C.I. Garcia and V. Thillou: Materials Science and Technology 2004 (AIST and TMS, New Orleans 2004), p. 3.

[239] L.K. Chiang: 4th Intern. Conf. on HSLA Steels, (Xi'an, China 2000).

[240] P. Uranga, B. López and J.M. Rodriguez-Ibabe: Proc. Second Baosteel Biennial Academic Conf. Vol. 2 (Shanghai, Baosteel 2006), p. 75.

[241] C.-P. Reip, W. Hennig, J. Kempken and R. Hagmann: Mater. Sci. Forum Vol. 500-501 (2005), p. 287.

[242] R. Zagatti: Thinner Slab Casting. 33rd McMaster Symp. on Iron and Steelmaking (McMaster Univ., Hamilton 2005), p. 301.

[243] R.J. Glodowski: 42nd Mechanical Working and Steel Proc. Conf. Vol. 38 (ISS, Toronto 2000), p. 441.

[244] W. Rohde: Rev. Métall.-CIT (1994), p. 529.

[245] H-P. Schmitz, H. Janssen and M. Bössler: 2nd Intern. Conf. on Thermomechanical Processing of Steels (Verlag Stahleisen, Düsseldorf 2004), p. 167.

[246] C.I. Garcia, C. Tokarz, C. Graham, M. Vazquez, L. Ruiz-Aparicio and A.J. DeArdo: Intern. Symp. on Thin Slab and Rolling (CSM, Guangzhou 2002), p. 194.

[247] G. Flemming and K.-E. Hensger: Thermomechanical Processing of Steels (IOM, London 2000), p. 537.

[248] C. Klinkenberg and K.-E. Hensger: Mater. Sci. Forum Vol. 500-501 (2005), p. 253.

[249] J.G. Williams, C.R. Killmore, F.J. Barbaro, A. Meta and L. Fletcher: Microalloying'95 (ISS, Pittsburgh 1995), p. 117.

[250] W. Ruizhen, Z. Hongtao, P. Ganyun, H. Yuchao, L. Ni and C. Guoki: Intern. Symp. on Thin Slab and Rolling (CSM, Guangzhou 2002), p. 292.

[251] X. Chuanfen, J. Liandi, X. Liqun, X. Zhiru, Y. Xuchang, L. Jiegeng and C. Xuewen: Intern. Symp. on Thin Slab and Rolling (CSM, Guangzhou 2002), p. 262.

[252] G.M. Megahed, S.K. Paul, T.A. El-Bitar and F. Ibrahim: Mater. Sci. Forum Vol. 500-501 (2005), p. 261.

[253] Dilihua: Intern. Symp. on Thin Slab and Rolling (CSM, Guangzhou 2002), p. 275.

[254] C.-P. Reip, W. Hennig, J. Kempken and R. Hagmann: Mater. Sci. Forum Vol. 500-501 (2005), p. 287.

[255] C.D. Horvath and J.R. Fekete: Int. Conf. on Advanced High Strength Sheet Steels for Automotive Applications (AIST, Winter Park 2004), p. 3.

[256] N. Pottore, N. Fonstein, I. Gupta and D. Bhattacharya: Developments in Sheet Products for Automotive Applications, MST'05 (Pittsburgh, MST 2005), p. 97.

[257] T. Heller, I. Heckelmann, T. Gerber and T.W. Schaumann: Recent Advances of Nb Containing Materials in Europe (Verlag Stahleisen, Düsseldorf 2005), p. 23.

[258] C. Bilgen, K.-E. Hensger and W. Hennig: Int. Conf. on Advanced High Strength Sheet Steels for Automotive Applications (AIST, Winter Park 2004), p. 141.

[259] W. Bleck, A. Frehn and J. Ohlert: Niobium Science & Technology (TMS, Orlando 2001), p. 727.

[260] M. Korchynsky: Microalloying'95 (ISS, Pittsburgh 1995), p. 449.